OUR ENIGMATIC UNIVERSE

One Astronomer's Reflections on the Human Condition

Alan H. Batten

Published by

MELROSE BOOKS

An Imprint of Melrose Press Limited
St Thomas Place, Ely
Cambridgeshire
CB7 4GG, UK
www.melrosebooks.com

FIRST EDITION

Copyright © Alan H. Batten 2011

The Author asserts his moral right to
be identified as the author of this work

Cover designed by Matt Stephens

ISBN 978 1 907732 03 4

All rights reserved. No part of this publication may be reproduced, stored in a retrieval system, or transmitted, in any form or by any means electronic, mechanical, photocopying, recording or otherwise, without the prior permission of the publishers.

This book is sold subject to the condition that it shall not, by way of trade or otherwise, be lent, re-sold, hired out or otherwise circulated without the publisher's prior consent in any form of binding or cover other than that in which it is published and without a similar condition including this condition being imposed on the subsequent purchaser.

Printed and bound in Great Britain by:
CPI Antony Rowe. Chippenham, Wiltshire

*In memory of Lois and
for Michael, Margaret, Terry, Erica,
Naomi, Alexandra, and Nicholas.*

PREFACE

In recent decades, historians of science have amply demonstrated that the popular notion that science and religion are necessarily opposed to each other is false, while several theologians have tried to take account of modern scientific knowledge in their theology. My impression, however, is that these efforts have only rarely come to the attention of either working scientists or members of religious congregations (including many of the clergy). At first sight, these two groups may appear to have little in common, but many of their members share an intelligent interest in the explorations of modern thought.

The conflicts often perceived between scientific discoveries and religious beliefs arise nowadays in many different subject areas: astronomy, biology, geology, neuroscience, and physics. No-one can be expert in so wide a range of subjects and, indeed, I do not even claim to be expert in all the areas of astronomy that I discuss. I hope to have avoided outright blunders, but I am uncomfortably aware that some readers will feel that I have either misunderstood, or oversimplified at least, matters about which they are knowledgeable. This may be especially true where I have ventured into philosophy and theology. I believe, however, that there is value in one person trying to integrate the results of these diverse lines of inquiry and that doing so is worth the risk of making mistakes. I am, perhaps, feeling my way to a new kind of natural theology. Since that kind of theology has been out of favour since the time of Charles Darwin, not only with scientists but also with many theologians, I will not be entirely surprised if I am assailed from both sides. I find this way of thinking helpful, however, and hope that it might prove so to others.

Progress in many of the fields that I have tried to cover is so rapid that what is up to date at the time of writing may well be out of date by the time that this book becomes available to readers. This is particularly true of my discussion of our knowledge of planetary systems other than our own. I have probably revised those few pages more often than I have any other part of the book but, almost certainly, they will be dated by the time you read them.

I am grateful to several friends and colleagues who have read through earlier versions of this work and have helped me to improve it: Harold Coward, Michael Hadley, John Leslie, Murdith MacLean and Terence Penelhum. My long association with the Centre for Studies in Religion and Society in the University of Victoria (Canada), under its three Directors, Harold Coward, Conrad Brunk and Paul Bramadat, has also helped me to develop my ideas, as did the year spent co-teaching a course on science and religion with Paul Wood of the Department of History of the University of Victoria. The usual proviso of course applies: any remaining errors and infelicities are my responsibility alone.

I am also grateful to Jill de Laat, Kathy Kimbrough, Paul Judges and others at Melrose Books for their help in turning my manuscript into a published book. The following acknowledgments are made for permission to reproduce previously published and copyrighted material: The Bertrand Russell Peace Foundation for the quotation from *A Free Man's Worship*; Basic Books (The Perseus Group) for the quotation from Dorothy Sayers' Inroduction to her translation of Dante's *Il Purgatorio*; Curtis Brown Inc. for a quotation that is also reprinted with the permission of Scribner, a Division of Simon & Schuster Inc., from THE ASTONISHING HYPOTHESIS by Frances Crick. Copyright © 1994 by the Francis H. C. Crick and Odile Crick Revocable Trust. All rights reserved. The quotation from Albert Einstein's *The World as I See It* is reprinted with permission from the Hebrew University of Jerusalem. I have been unable to trace the current owners of the rights to Eddington's *Science and the Unseen World*.

Victoria, B.C. Canada
August, 2010

LIST OF CONTENTS

Preface v

Prologue ix

Chapter 1: Humanity's Expanding Perception of the Universe 1

Chapter 2: How We Perceive the Universe 17

Chapter 3: Belief in God 37

Chapter 4: The Argument from Design 51

Chapter 5: Life and Consciousness in the Universe 77

Chapter 6: The Mind and the Brain 91

Chapter 7: Reason and Revelation 116

Chapter 8: Imagery, Miracles and Prayer 131

Chapter 9: The Historical Relations of Science and Religion 149

Chapter 10: The Harmonization of Science and Religion 165

Epilogue 191

Name Index 197
Subject Index 203

PROLOGUE

> Our birth is but a sleep and a forgetting;
> The Soul that rises with us, our life's Star,
> Hath had elsewhere its setting
> And cometh from afar:
> William Wordsworth, *Intimations of Immortality*

At the moment of birth, each one of us begins a quest: to explore the world into which we have been thrust and to try to understand our own relationship to it. There is an outer aspect to our world, the physical one, which, as we grow, becomes more familiar, and some of us learn to study it by the methods of the natural sciences; but there is also an inner aspect to the world, consisting of our thoughts, emotions and values, that does not easily lend itself to investigation by the same means. Traditionally, that inner world has been the sphere of the world's great religions, but many of the most successful interpreters of the findings of modern science argue strongly that religious belief and scientific knowledge are incompatible, or even that religious belief has been positively harmful to the human race. Many of these writers are convinced that, in time, the study of the human brain will enable this inner aspect of our world to be explained by the methods of natural science. On the other hand, historians of science have, in the last two decades or so, shown how complex is the interrelation between the scientific and religious methods of exploration. Many, if not most, of those whom we now regard as the founders of modern Western science, owed their inspiration to their religious convictions. Has the progress of science shown that that inspiration was false, or is it still possible to hold a meaningful religious belief while accepting the results of modern science? There are strong voices on both sides urging us to choose one or the other path of exploration, claiming that the two options are mutually exclusive, but perhaps each path has the potential to enrich our lives in its own way.

Whatever subconscious memories we may retain of our time within the womb, birth is in a real sense a beginning for each of us in our individual quests, so Wordsworth's claim that birth is a sleep and a forgetting seems somewhat paradoxical. In fact, modern science and orthodox Christianity would agree in dismissing that claim as erroneous, although some other religions would be more open to it. Whatever the truth may be, we did not choose to be born; our coming into this world is the result of the actions of two people who do much to determine what the courses of our several lives shall be; biologically they pass on their genes to us, and culturally, at least for most of us, they control our earliest and formative experiences. They themselves came into the world and became what they were through a similar train of events. Our lives seem to be the result of a series of accidents and we wonder about our own significance. Yet, with possible exceptions for those who suffer great pain from medically incurable diseases, most of us think it wrong even to try to end our lives before they have run their respective natural courses.

None of us can remember how we first began to make sense of this world into which we have been plunged, but those of us who have become parents ourselves can make some reasonable guesses about what it must have been like. The baby has first to learn that it is distinct from the world, that there are other people in that world who are also distinct from it, and, most important of all, that the world does not revolve around the baby nor exist solely for the latter's benefit. To judge by the behaviour of some adults, some babies never learn this last lesson – and perhaps all of us are guilty of forgetting it on occasion. There is still one more lesson to come: we have to come to terms with the fact that, like all who have gone before us, we must die.

There are some parallels between our individual experiences and the collective experience of the whole human race. We probably never will know just at what stage in their development human beings, or their immediate ancestors, became fully aware of themselves, of the world around them, including the objects that they could see in the sky, and of the difference between right and wrong. We can guess, however, that the immediate ancestors of those who were, in that sense, the first to be fully human, were just as self-centred as newborn babies still are. They, too, had to learn to distinguish themselves from the world, to respect each other, and to try to understand their own relationship with the world and with each other. After all, commonsense observation seemed to show that the world (in its largest sense of the cosmos) *did* revolve around them. It was at first quite reasonable to suppose that the world, and everything in it, was made for the benefit of the human race. Our early ancestors soon learned to domesticate animals and to bend those creatures to their own will, and to make use of many plants. Even the Sun to light the day and the Moon the night were seen to serve human purposes, and observation again led our ancestors to believe that we were at the centre of everything. It was natural to suppose that a beneficent Deity had created us and this world for us to live in.

Like babies, we have had to learn that the world is not so simple. Learning that took us longer than it takes most modern babies (at least in the developed world) and not much progress was made in our understanding of the *physical* universe until the rise of modern science, some four centuries ago. Of course there was what we now call "science" before that – in ancient Greece, in India and China, and in the Moslem Caliphate – but something different in Europe, in the late sixteenth and early seventeenth centuries, turned natural science into a steadily progressive accumulation of knowledge about the physical world which has radically changed our view of our relationship with that world and often runs counter to common sense. In the past century, scientific research has proceeded at a dizzying pace, so that even those of us who have devoted our lives to its pursuit have a hard job to keep up with all developments, even in our own field of work, let alone the sciences in which we are not expert. Many of our number cannot understand why some people reject this new knowledge and cling to traditional expressions of our relationship to the world, enshrined in writings that are held to have a sacred status; perhaps those scientists should remind themselves just how new their knowledge is. Modern science tells us that we are not at the centre of creation – there may be no such place – that we are akin to the other animals, and now even our brains are being probed and our thoughts and hopes and wishes being reduced, in the words of Francis Crick[1], to "nothing but a pack of neurons" – a point that we shall discuss in Chapter 6. Wordsworth's poetic vision of our having come from afar, trailing clouds of glory, always suspect in the Christian West, appears to have been completely ruled out by modern science. According to the outlook of some modern scientists, nothing that can be called "you" existed before your parents copulated (or indeed, many would now say, not until some time after that event) and, whatever you are, you will cease to exist within about ten minutes, at most, after your heart and breathing stop.

Francis Crick was only one of many scientific authors who proclaim this view as the inevitable conclusion to which modern science leads us: but is it so inevitable? Science tells us a lot about how the physical aspect of the world works, and scientists have done much to make our lives in that world easier and more comfortable. While many have consequently turned their backs on the possibility that the physical world may be only a part of all that there is, not everyone who accepts the findings of modern science would necessarily subscribe to Crick's extreme form of materialism, which it is one of my aims to challenge. Perhaps it is true that there is nothing beyond the physical, but it is no more possible to prove that so than it is to prove the opposite. We may yet ask whether that particular kind of naturalism is the only rational response to all our new knowledge, or if the ancient religions of the world still have something to teach us today and are not just unfortunate and harmful relics of the infancy of our race, as writers like Francis Crick, Richard Dawkins,

Carl Sagan, Steven Weinberg and many others try to persuade us. They are or were all highly skilled scientists and deservedly popular writers, but that does not guarantee that they are right about everything. Curiously, their message is scarcely changed from that written by Bertrand Russell[2] over a century ago, in *A Free Man's Worship*:

> That Man is the product of causes that had no prevision of the end they were achieving; that his origin, his growth, his hopes and fears, his loves and his beliefs, are but the outcome of accidental collocations of atoms; that no fire, no heroism, no intensity of thought or feeling, can preserve an individual life beyond the grave; that all the labours of the ages, all the devotion, all the inspiration, all the noonday brightness of human genius, are destined to extinction in the vast death of the solar system, and that the whole temple of Man's achievement must inevitably be buried beneath the debris of a universe in ruins – all these things, if not quite beyond dispute, are yet so nearly certain, that no philosophy which rejects them can hope to stand. Only within the scaffolding of these truths, only on the firm foundation of unyielding despair, can the soul's habitation henceforth be safely built.

If Russell and his modern disciples are right, then our life's quest is a sorry affair indeed; yet although more than a century has elapsed since Russell wrote those words and we know much more about the structure of the universe and the workings of the human brain than he did then, philosophies that reject his near certainties are still being entertained. We cannot deny the evidence, but perhaps we may look at it in a different way from that of the materialistic authors I have named. Indeed, there are good reasons for questioning the very first clause in the first sentence of the above quotation, as we shall see in Chapter 4. In the pages that follow, I shall attempt to look at the evidence in a way different from Russell's, and to present a rational case for an outlook that can be called religious, even if it fits only awkwardly into the framework of Christian orthodoxy. Three themes, therefore, will permeate the book: that materialism is not the only tenable philosophy for those who take seriously the discoveries of modern science about the size, age and nature of the universe; that a rational case can be made for some form of religious belief; and that, despite all the astounding progress of recent scientific discovery, the universe is indeed enigmatic and there is still much to be learned about the vastness of space, the unimaginably long stretches of time, and still more about the complexity of our human nature and personalities.

References:

[1] Crick, F.C., *The Astonishing Hypothesis* (Simon and Schuster, London and New York, 1996), p. 3.

[2] Russell, B., *A Free Man's Worship*, 1903, reprinted in *The Basic Writings of Bertrand Russell* (Simon and Schuster (Touchstone Books), New York, 1961), eds. R.E. Egner and L. Denonn. (George Allen and Unwin Ltd, London), pp. 66-72.

CHAPTER 1
Humanity's Expanding Perception of the Universe

> Le silence éternel de ces espaces infinis m'effraie.
>
> Blaise Pascal, *Pensées*

Science has its fashions and they are often reflected in the titles of popular books on scientific subjects. When I was a young boy, avid to read everything on astronomy that I could find, most of the books that came my way had the word "stars" in their titles. There were exceptions of course; Sir James Jeans[1,2] wrote *The Universe Around Us* in 1929 and *The Mysterious Universe* in 1930. Sir Arthur Eddington[3] followed with *The Expanding Universe* in 1933. Those two giants were ahead of their time, however; it took a generation or two for other science writers to catch up. At the present time, you can hardly find a popular book on astronomy that does *not* contain the word "universe" in its title unless, perhaps, it has the word "cosmos" instead. This change in the titles of popular books reflects faithfully the change in the focus of interest of the majority of professional astronomers. Although I could not know it, that change had already begun while I was devouring books about "the stars". Popular books, especially those written for young children, are often behind the frontiers of research, and the time lag was greater in the Britain of World War II, where I spent my childhood. Edwin Hubble's[4] discovery in 1929 of the law relating the velocities of recession of the spiral nebulae (as other galaxies were then known) to their distances, and Einstein's[5] 1915 theory of general relativity are the most prominent markers of the beginning of modern scientific cosmology; but the same war that delayed the popularization of science also delayed the pursuit of disinterested scientific research. Scientific cosmology did not really unfold until that war was over and our own little world had returned to some semblance of normality.

Before those landmark discoveries mentioned in the last paragraph, the word "cosmology" denoted a branch of philosophy rather than of science. People did not think that astronomers could tell us very much about the universe as a whole, or of our relationship to it. Cosmology as a science is not much older than the oldest persons still living, and is younger than my own parents would be if they were still alive. Indeed, astronomers of my parents' generation, many of whom I knew, began their careers when it was still uncertain whether what were then called the "spiral nebulae" were other galaxies, or a special class of objects within our own Galaxy, the Milky Way. Throughout the nineteenth century, most astronomers believed the universe to be limited to the system of stars that we see every clear night. If that system should prove to be finite, and Olbers' famous paradox appeared to show that it must be, there was, presumably, nothing but empty space beyond.* Although Kant[6] had introduced the idea of "island universes" and the discovery – made in 1845 with the great telescope built in Ireland by Lord Rosse – of the spiral structure of some "nebulae" was seen by others as a possible confirmation, the idea remained speculation until after Hubble's discovery. Even Hubble himself

---

* The paradox, associated with the German astronomer H.W.M. Olbers (1758-1840), is that if the universe were infinite with stars distributed uniformly throughout it, the night sky should be uniformly bright since in every direction we would eventually see a star. The Milky Way is an approximation to what the whole sky should look like in such a universe. In modern terms we would substitute galaxies for stars. Since the night sky is dark, it was argued, the universe must be finite. For some time, it was believed that the paradox was resolved by the expansion of the universe and that an expanding universe could be infinite. It is now recognized that the fact that stars do not shine for ever provides the required resolution.

was reluctant to accept the expansion of the universe that he is now credited with having discovered.

Scientific cosmology could not develop before the early twentieth century because its study required large telescopes. Even the telescope of six-feet (1.8-m) aperture built by Lord Rosse, and the 100-inch (2.5-m) that Hubble used could give only tantalizing glimpses of what might be waiting to be discovered in the distant reaches of the universe. The Great Depression of the 1930s and World War II delayed the construction of larger instruments. When the 5-m telescope of Palomar Mountain was commissioned in 1948, it was believed to be about as large an optical telescope as could be made. Although Soviet Russia followed with a 6-m telescope, its builders encountered many difficulties that seemed to confirm that the limit had been reached. More recent advances in technology, however, have enabled the construction of 8-m and 10-m telescopes, which are now in operation, and plans for telescopes of 30-m aperture, or even larger, are being made. The commissioning of the Palomar telescope coincided with the return from war work to academic research of many brilliant scientists, and a consequent ferment of both observational and theoretical activity in astronomy. In the late 1940s, two rival theoretical cosmologies emerged. The first, now called "Big-Bang cosmology", is often associated with the name of George Gamow[7], who published a brief paper on it in 1946, followed by more detailed ones later (with Hermann, Alpher, and Hans Bethe), although Georges Lemaître in Belgium and Alexander Friedmann in Russia had anticipated many of Gamow's ideas before World War II began. (Gamow, in fact, had briefly studied under Friedmann.) The second cosmology, the "Steady-State theory", particularly associated with the names of Fred Hoyle, Hermann Bondi and Thomas Gold – all three only recently dead – was put forward at about the same time. Developments since the mid-1960s have convinced most astronomers that the Big-Bang theory is correct in its essentials, but the consensus is not complete. While all agree that the original form of the Steady-State theory is untenable, some still hope that a more complex form of it will explain the observations. Considering the comparative youth of scientific cosmology, we would perhaps be unwise to assume that *any* of our current theories will stand after another century. The universe, as J.B.S. Haldane[8] once put it in a remark to which we shall refer again, may well be not only queerer than we suppose, but queerer than we *can* suppose.

That there is queerness aplenty in the physical universe is another discovery of twentieth-century astronomy. Just as important as the increases in the size of our ground-based telescopes, if not more so, has been our ability to send quite modest instruments above the atmosphere of the Earth. From that vantage point, we can detect types of radiation that do not penetrate our atmosphere to reach us on the ground, and that have revealed to us a whole "zoo" of cosmic objects that few of us dreamt of before they were discovered. Some years ago, a colleague who had been a graduate student with me, speaking to a professional audience, reminded the older ones among us that, if we had been told in our student years what would be the principal objects of interest to astronomers at the end of the twentieth century, not only would we not have believed the list, we would have been unable even to understand it! He was right; "quasar", "pulsar" and "black hole" are all terms coined since I received my doctorate. While there were theoretical grounds for searching for what are now called black holes, the other two terms were coined for objects that were complete surprises. Terms like "dark matter" and "dark energy" have also been coined to describe things that we do not fully understand. Detailed observations of galaxies have convinced many that there is more matter in the universe than we can see, and that it must be different from any of the known fundamental particles. Recent observations have also shown that the expansion of the universe seems to be accelerating – as if energy were still being poured into the system. The universe is mysterious in many more ways than Sir James Jeans imagined, and my guess is that the larger telescopes now being planned will bring at least as many new surprises as elucidations of old puzzles.

Nevertheless, it is the vastness of the universe revealed by modern cosmology that seems to us, now, to be its most salient characteristic. The expansion of the universe was itself an unexpected and dramatic discovery, but, in some ways, the expansion of our *concept* of the universe is even more dramatic. While the twentieth century was a time of particularly rapid change in our concepts, the history of astronomy since the time of Copernicus has been, in large measure, the history of our expanding perceptions of the universe of which we are a part. Astronomers brought up in the Ptolemaic system, in which Sun, Moon, planets and fixed stars all revolve around the Earth, thought that they knew the size of the universe. They did know the distance from the Earth to the Moon fairly accurately. The distances to the other planets (in their terminology, both Sun and Moon were planets) were worked out on the reasonable assumption that the crystalline spheres believed to carry each planet nested neatly, one inside the other. The outermost sphere, also centred on the Earth, carried the fixed stars, all at much the same distance, which was believed to be somewhat less than the modern value for the distance of the Earth from the Sun. The whole was driven by Aristotle's *primum mobile* and beyond that, at least in the Christianized version that dominated European thought until the time of Galileo, was the Empyrean, or the abode of God.

The Ptolemaic universe was a comfortable cosmos, a user-friendly universe. Early in the fourteenth century Dante[9], well-versed in Ptolemaic astronomy, could imagine traversing the whole universe, from the centre of the Earth to the Empyrean, in about a week, even though he stopped for very many lengthy conversations on the way. Admittedly, Dante supposed himself to have been given supernatural aid, but the timescale of the journey would not have seemed completely incredible to his contemporaries, who would, of course, have clearly recognized its imaginary nature. Copernicus changed all that, because both he and those who continued to adhere to the Ptolemaic system recognized that, if the fixed stars were as close to the Earth as the latter believed, they should show the annual changes in their relative positions that are known as stellar parallax. No such parallaxes could be observed at the time of Copernicus, nor would they be observed until nearly three centuries later. This was the strongest argument against the Copernican hypothesis, and its strength should not be underestimated by those who wonder why so many took so long to accept that hypothesis.

Copernicans, including Copernicus himself, who died before the invention of the telescope, were quick to respond with what was eventually demonstrated to be the correct answer: the fixed stars are so far away that their parallaxes could not be detected with the instruments available at that time. The argument was correct, but it must have seemed like an ad hoc assumption to those wedded to the older hypothesis. By displacing the Earth from the centre, Copernicus also struck at the heart of Aristotelian physics, without offering any physical hypotheses of his own – they were to come from Galileo, Kepler, Descartes and above all, Newton. Going against the current physical theories and dismissing a strong counter-argument to the Copernican hypothesis by claiming that the fixed stars were many times farther away than had previously been dreamt was not calculated to win over conservative philosophers, let alone theologians who had come to have a vested interest in a geocentric universe.

There were, however, problems for theology implicit in the Copernican hypothesis that were even greater, at least for those inclined to literal interpretations, than the displacing of the Earth, and therefore the human race, from the centre of the universe. In Ptolemaic cosmology, the fixed stars were all disposed in a relatively thin shell so that there were no great differences in their distances from the Earth. As we have seen, in the Christianized version of that cosmology, beyond this shell was the abode of God. With the Sun in the centre and the stars moved to very great distances, it was inevitable that, before long, someone would propose that the distances of the individual stars might differ very considerably from each other, occupying a very large and possibly infinite volume. Where then would be the abode of God? Copernicus (1473-1543) did not take that step himself, but the Englishman Thomas Digges[10] (1545-1595), and more famously the Italian Giordano Bruno (c. 1548-1600) living only a little later, did (and Bruno took the further step of supposing that there could be many other planetary systems containing habitable planets).

Of course, the best theologians do not and did not think of God being located in space, but the imagery of God beyond the sphere of the fixed stars was employed by Dante and was probably taken literally by many of his contemporaries. Moreover, a universe infinite in space might also have existed, and continue to exist, for an infinite time – and, on the face of it, that might seem to contradict the doctrine of creation *ex nihilo* which, since the finalization of the Nicene Creed in the fifth century, has been an important part of orthodox Christianity. Aquinas grappled with this problem in his attempts to reconcile Aristotelianism with Christianity and concluded that "creation" did not necessarily imply a temporal relationship: an eternal universe, such as Aristotle believed in, could still be completely dependent on God for its continued existence. Many modern theologians follow this line of thinking and Helge Kragh[11], a Danish historian of science, has stressed several times in a recent book that the statement "God created the universe" need not imply that God existed before the universe. Undoubtedly, however, some Christians at some times have been troubled by the idea of an infinite and eternal universe. When Bishop Tempier of Paris condemned a number of propositions in 1277, among them were several that clearly implied the Aristotelian eternal universe. Ironically, however, the first to resist the new cosmology of the sixteenth and seventeenth centuries were not orthodox Christians, but the so-called Aristotelian professors of philosophy in the universities. Only as the implications of the new cosmology for the ordinary believers' estimate of their worth in relation to the universe began to sink in, were Christians disturbed enough to join the opposition.

As everyone knows, the confrontation between the two world-views came to a head during the life of Galileo (1564-1642). Galileo's trial and the events surrounding the publication of Charles Darwin's[12] *Origin of Species* (1849) have become archetypal examples of a supposed "conflict" between science and religion, although historians of science have marshalled ample evidence that both these episodes were much more complex than is often supposed. As already mentioned, Galileo's quarrel was at first not with the Church, but with the Aristotelian professors of philosophy – particularly of natural philosophy. The new cosmology undermined Aristotelian physics because within it the centre of the Earth no longer coincided with the centre of the universe; therefore the assumption that heavy objects fall to the centre because that is their natural place no longer held. On this account, Copernicanism would have met resistance even if metaphysical and religious issues had not got entangled with those that we would term purely scientific. Aristotelians, however, began to buttress their physical arguments with scriptural ones, particularly appealing to the miracle of Joshua, who commanded the Sun, *not* the Earth, to stand still. Galileo would have been wiser to ignore that argument, but he could not resist the temptation to advance his own interpretation of Scripture (recently admitted by Pope John Paul II to be quite in accord with modern Roman Catholic exegesis) and thus moved on to what was, in seventeenth-century Italy, very dangerous ground. Shaken by the Protestant Reformation and involved in the Thirty Years War, the last thing the Church hierarchy needed was an internationally renowned layman teaching it how to interpret the Bible.

Undoubtedly, the personalities involved, not least Galileo's own, had a great deal to do with the final outcome, but at the heart of the encounter was the change from a relatively small comfortable cosmos, created specially for the human race, to a universe that was certainly much larger and possibly even infinite in space and time, in which the significance of the human race was not immediately obvious. Europeans, taught to regard human beings as having been made "in the image and likeness of God", understandably found this radical change in thought difficult and disturbing. Many people have not fully adjusted to it even yet. Galileo's younger contemporary, Blaise Pascal, summed up the import of the change in the words quoted at the head of this chapter, which have become famous and can be rendered into English as: "The eternal silence of those infinite spaces terrifies me." In our own time, some have gone even further, denying all meaning to the cosmos and to our own existence. Steven Weinberg[13], Nobel Laureate in physics, summed up *that* attitude in an oft-quoted remark: "…the more the universe seems comprehensible, the more it also seems purposeless." We can only wonder if Copernicus[14], a churchman, foresaw even dimly how revolutionary his *De Revolutionibus* would turn out to be. I incline, however, to the opinion that his successors in the Church hierarchy, the contemporaries of Galileo, did have some inkling of what was to come and, in their own eyes, that fully justified them in trying to suppress public dissemination as fact of a theory that, at that time, neither had been, nor could be, verified observationally.

While astronomers set in motion the processes that led to the great change in our attitude to the universe, they did not effect the change all by themselves. The next to challenge received opinions, in the late eighteenth and early nineteenth centuries, were the early geologists, particularly Hutton and Lyell. Just as Copernicus, Kepler, Galileo and Newton had opened up vast vistas of space for contemplation, Hutton and Lyell did the same for time. Although, as we have seen, there was the implied possibility of an eternal universe within the Copernican scheme, few in Christendom had questioned that the Earth was more than a few thousand years old until the late eighteenth century. Archbishop Ussher's famous date for the Creation, 4004 BCE, is well known through having been enshrined in the margin of the King James Version of the Bible, but it was only one of many attempts to work backwards, taking the biblical evidence at its face value. The evidence does not admit of a unique answer, but all scholars who tried to work out a date for Creation found the age of the Earth to be only a few thousand years – a value that the more extreme creationists still insist on. In fact, as modern archaeology has shown, ten thousand years is not a bad estimate of the age of human civilization; and for those willing to accept that the early chapters of *Genesis* contain folk memories of those very early times, from which, nevertheless, we may gain useful insights into the relationship between the human race and the rest of creation, the difference between the biblical and scientific timescales is simply not an issue. On the other hand, those who believe that even the implications of the biblical narrative are part of the infallible word of God (the Bible nowhere gives an *explicit* date for the Creation) the great age of the Earth posited by modern science creates a problem that many of them can solve only by rejecting the science.

This expansion in our concept of the *age* of the universe has been, if anything, more troubling to those who believe in the literal truth of the Bible than was the earlier expansion of our concept of the *size* of the universe and its concomitant displacement of the Earth from the central position. Very few people nowadays wish to argue on the strength of Joshua's alleged miracle that the Earth must be stationary at the centre of a relatively small cosmos, but there are still many who want to argue that the Earth cannot be much more than 10,000 years old, and probably must be even younger, and that the human race is correspondingly young and was specially created.

That last point, of course, is the rub. The geologists prepared the way for Darwin and his theory of evolution; they gave him the long periods of time needed for the process of natural selection to be effective. The human race was not merely reduced to a tiny speck in the vast expanse of space and to a mere episode in the equally vast expanse of time; it was shown, although Darwin did not labour the point, in the *Origin* at least, to be closely related to the rest of the animal creation and, as many now believe, an accidental by-product of an impersonal process. It is natural for those who balk at such conclusions to insist on a young Earth; they are well aware that nothing would undermine Darwinian evolution so completely as a clear demonstration that the Earth cannot have existed for the long periods of time that astronomers, biologists and geologists now agree on.

The biblical literalists are fighting a lost battle but, reflecting that within the last two centuries the learned world has increased our estimate of the age of the Earth by a factor of nearly a million, and the age of the universe by still more, we should not be all that surprised that many people find it difficult to accept the new estimates of these ages. Moreover, through much of the nineteenth century a *scientific* debate raged about the age of the Earth. The great physicist Lord Kelvin (1824-1907), throughout his long life, vigorously opposed the uniformitarian geology of Hutton and Lyell (see Burchfield[15]). Kelvin's motives were not primarily religious, although he did think it important to demonstrate that the Earth (and the universe) must have had a beginning. Kelvin simply believed that the geologists were ignoring some basic physics; his shots were aimed at Hutton and Lyell, before ever Darwin published the *Origin of Species*. Kelvin was not opposed to the idea of evolution itself, although he did have some reservations about the process of natural selection. The understanding of the physical principles on which he relied, primarily the theory of heat, was changing and growing throughout the nineteenth century and empirical measurements were also being improved, so Kelvin returned again and again to his analyses, especially to the one based on the rate of cooling of the Earth, in an effort to refine his calculations and to give a definitive answer. Inevitably, therefore, Darwin and evolution were caught in the crossfire. Coupled with biological arguments, mainly arising from nineteenth-century ignorance of the mechanism of heredity, Kelvin's attacks gave fuel to religious opponents of Darwinism, even though many churchmen were open to the new ideas. No more than at the time of Galileo was there a simple battle between science and religion. Support of, and opposition to, both Copernicanism and Darwinism cut across any divide between believers and atheists.

Kelvin's arguments, of which there were three, seemed strong enough during most of the nineteenth century. The first argument was that the Earth did not have a sufficient store of internal heat for it to have been warm enough to sustain life over the long periods of time of which the geologists talked. The second argument was that the Sun itself could not supply heat and light for that long. The third argument was that, over the long periods of time that the geologists believed the Earth had existed, the tidal effects of the Moon would have slowed down the Earth's rotation until the Earth always showed the same face to the Moon, as the Moon does to the Earth. The first argument is usually represented as having been invalidated by the discovery of radioactivity in the last decade of the nineteenth century, and Rutherford made this point in 1904 (in Kelvin's lifetime and in his presence). The true story, once again, is more complex, as shown in a recent paper by Philip England[16] and others. Although the argument from radioactivity seems to have been the one that carried conviction, a decade before Rutherford, John Perry had pointed out that if the Earth had a molten convective core, its surface could be kept heated very much longer than Kelvin's computations suggested. The third argument was criticized, also in Kelvin's lifetime, appropriately enough by one of Darwin's sons, the famous mathematical astronomer Sir George Darwin. Although the tidal effect invoked by Kelvin is real, the calculations he based on it were at fault. Only the second argument survived its originator but even so, three decades after Kelvin's death a source of energy sufficient to keep the Sun and other stars shining for billions of years was identified, namely thermonuclear fusion.

Right at the beginning of the twentieth century, the great Dutch botanist Hugo de Vries[17] was still influenced by Kelvin's arguments and supposed that at most twenty to thirty million years were available for evolution. He was one of those who rediscovered Mendel's laws and he introduced the term *mutation* into evolutionary biology with something like its modern meaning. Mutations, contrasted with natural variability such as he supposed Darwin had had in mind, were necessary to bring about variations on which natural selection could work in such limited time as de Vries believed to be available. His ideas contributed to the overcoming of biological objections to Darwinian evolution and the emergence of the so-called "modern synthesis" of genetics and evolutionary science just as objections based on physics were also seen to have lost their force. What we now call DNA was first isolated from a cell in 1869, only a few years after Mendel's work was published; that molecule was identified as the carrier of heredity in the early 1940s and then, as is well known, its structure was elucidated in 1953 by Watson and Crick, revealing the means by which mutations could occur and natural selection could act. So it was just about a century after the publication of the *Origin of Species* that scientific objections to Darwinism were removed and opposition to evolution by natural selection became a purely religious matter. Even at the time of the Scopes trial in 1925, scientists could still be found who had reservations about that form of evolution.

The fate of Kelvin's second argument underlines again the rapidity of the changes in our conception of the universe. Astronomers were feeling their way toward a sub-atomic source for stellar energy from about 1920 onwards, and conclusive demonstration that the conversion of hydrogen into helium, deep inside the stars, could provide the energy needed for the length of time that both geologists and evolutionary biologists required, arrived on the scene in 1938, within my own lifetime. Until then, any fundamentalist could have argued (and, for all I know, some did) that astronomers did not know the origin of stellar energy and therefore the universe might be quite young – just as now some argue against the theory of evolution because biologists do not know the origin of life. I will come to the dangers of basing arguments on things that scientists do not know in another chapter; the most important point here is that the list of things that scientists do not know changes very rapidly.

There is an opposite danger: religious believers may be tempted to climb too quickly onto a scientific bandwagon that appears to offer support for their beliefs. No less a person than St Augustine warned the Church against this, though many believe that he fell into that very trap himself. Certainly, as we have seen, the mediaeval Church tied its doctrines too closely to the Ptolemaic cosmology. Some modern Christians, even including a former Pope, have been tempted to see in Big-Bang cosmology an echo of the opening verses of *Genesis*; but Big-Bang cosmology may yet go the way of all other cosmological theories and will almost certainly be modified in the decades to come. As we have seen, there are still some who think that the data may be squared with a modified Steady-State theory, in which the universe lasts for ever and stretches through infinite space. While such a conclusion might be less acceptable to some Christians, Buddhists and Hindus would have no problem with an indefinitely long timescale, although they would envision it as populated by a series of universes. Oddly enough, even in Western science there are now some who speak of the possibility of many "universes" existing either concurrently or consecutively, an idea that I shall discuss in more detail in Chapter 4. Eastern religious thought and some strains of Western scientific thought, therefore, agree that we may well be part of an ensemble of universes that could be infinite both in space and time. So much for Kelvin's attempts to show that there must have been a beginning! Yet, only a few decades ago, the orthodox picture of the universe was that it was "finite but unbounded": no matter how far you travelled within it, you would never come to an edge and must eventually return to your starting point.

Perhaps it is the endlessness, or endless repetition, of the universe that people like Steven Weinberg find so meaningless. Half a century before Weinberg, Sir Arthur Eddington voiced the same sentiment but drew a different conclusion. Rejecting the idea of consecutive cycles of expanding and contracting universes (the idea of concurrent multiple universes had not then been formulated) Eddington[18] wrote: "It seems rather stupid to keep doing the same thing over and over again." Unlike Weinberg, Eddington, a Quaker, had a strong religious belief and did not think the universe to be either purposeless or stupid. Thus Eddington rejected the idea of multiple universes, for which, in any case, there was little or no evidence in his day. In our time, as we have just seen, the idea has been revived; one possible interpretation of quantum theory requires the existence of very many parallel universes (or worlds). Although this interpretation is favoured by many workers in the field, it is hardly proven fact. Indeed, it is hard to see how the existence of parallel universes could be demonstrated, even though it is, apparently, theoretically possible for information to pass from one to another.

Even if the existence of parallel universes is granted, does it necessarily follow that Weinberg is right and the whole ensemble is without purpose? Some would suppose that if the whole universe is purposeless then our individual lives must be also. If that were so, then our little lives would seem pointless and, after we have strutted and fretted our hour upon the stage, they will be rounded in a sleep – if I may be forgiven the conflation of lines from two of Shakespeare's plays! Others would argue that even within a universe that is itself purposeless, our lives can have meaning and it is incumbent upon us to live them as well as we can. If that is true, then does not the fact that this universe has brought into existence beings that can envisage purpose lead us to question whether, after all, it is itself purposeless? An implicit assumption in reaching the conclusion that the universe is without purpose is that the physical universe that I have been describing in this chapter is all that there is. Many would see that as an entirely reasonable assumption but I shall discuss its validity in more detail in the next chapter. I simply note here that, even if the assumption is valid, there is still much about the physical universe that we do not understand.

References:

[1] Jeans, J.H., *The Universe Around Us* (Cambridge University Press, 1929).
[2] Jeans, J.H., *The Mysterious Universe* (Cambridge University Press, 1930).
[3] Eddington, A.S., *The Expanding Universe* (Cambridge University Press, 1933).
[4] Hubble, E.P., *A Relation between Distance and Radial Velocity among the Extra-Galactic Nebulae, Proc. Nat. Acad, Sci.*, **15**, 168-173, 1929.
[5] Einstein, A., 1915, *Sitzungsberichte, Preussische Akademie der Wissenschaften*, p. 844.
[6] Kant, I., 1755, *Allgemeine Naturgeschicte mit Theorie des Himmels*, English translation by W. Hastie, 1900, reprinted as *Universal Natural History and Theory of the Heavens by Immanuel Kant*, with an introduction by Milton K. Munitz (Ann Arbor Paperbacks, Michigan University Press, 1969), p. 62.
[7] Gamow, G., *The Expanding Universe and the Origin of the Elements, Physical Review*, (2), **70**, 572-573, 1946.
[8] Haldane, J.B.S., *Possible Worlds and Other Essays* (Chatto & Windus, London, 1927), esp. pp. 260-286.
[9] Dante Alighieri, *La Divina Commedia*, 1314-1321.
[10] Digges, T., *A perfit description of the celestiall orbes*, in *Prognostications Everlastinge,* by Leonard Digges (Thomas Marsh, London, 1576).

[11] Kragh, H., *Matter and Spirit in the Universe: Scientific and Religious Preludes to Modern Cosmology* (Imperial College Press, London, 2007), pp. 342-349.

[12] Darwin, C.R., *The Origin of Species* (John Murray, London, 1859). Many reprints.

[13] Weinberg, S., *The First Three Minutes* (Basic Books Inc., New York, 1977), p. 154.

[14] Copernicus, N., *De Revolutionibus Orbium Coelestiun*, 1572.

[15] Burchfield, J.D., *Kelvin and the Age of the Earth* (Science History Publications, New York, 1985).

[16] England, P.C., Molnar, P., and Richter, F.M., *Kelvin, Perry and the Age of the Earth*, in *American Scientist,* Vol. 95, July-August 2007, pp. 342-349.

[17] de Vries, H., 1901, *Die Mutationstheorie: Versuchen und Beobachtungen über die Entstehung von Arten im Pflanzenreich*, Leipzig, Verlag von Veit & Comp., English translation by J.B. Farmer and A.D. Darbishire in 2 vols., *The Mutation Theory: Experiments and Observations on the Origin of Species in the Vegetable Kingdom* (Open Court Publishing Company, Chicago, 1909-1910), esp. the final chapter of Vol. 2.

[18] Eddington, A.S., *The Nature of the Physical World* (Cambridge University Press, 1929), p. 86.

CHAPTER 2

How We Perceive the Universe

> Now my own suspicion is that the universe is not only queerer
> than we suppose, but queerer than we *can* suppose.
> J.B.S. Haldane, *Possible Worlds and other Essays*

In cosmological discussions the words "world" and "universe" are often used interchangeably. In our childhoods, our universe *was* the world in which each one of us lived; indeed, just that very small portion of the world (in the sense of the globe of the Earth) that came under our immediate observation. As we saw in the last chapter, we even had to learn that, in some sense, that world was separate from ourselves. As we eventually became aware of Sun, Moon and stars, they seemed to be parts of the world, or perhaps attendant satellites of it – not that we could formulate the idea in such words! Presumably, this childhood experience, which most of us can barely remember, reflects the collective experience of the human race as its consciousness evolved to the level that has now been reached. Again, as we saw in the last chapter, our perception of the world, or universe, has been an expanding one for at least as long as the historical record is available.

Most of us are endowed with five senses with which to experience the world around us: sight, hearing, smell, touch and taste. The last two depend on actual physical contact with an external object, and hearing and smell are limited in range and dependent on the existence of a medium between us and the object, which conveys the appropriate stimuli. When we turn our attention to the "world" beyond the Earth – the universe – we are completely dependent on the sense of sight. Even when we are examining signals conveyed in parts of the electromagnetic spectrum that our eyes cannot detect, we transform them into some form of record that we can see.

Somehow, by stages that I suppose none of us can remember, we come to trust the information that our senses give us. In our immediate surroundings, the congruence of the different senses must have a lot to do with building up this trust. We experience the world as one of sights, sounds, smells, touches and tastes, and there are relationships between these different kinds of experience. Tastes and smells are particularly closely related, and things that glow usually feel hot – at least they did before the LED was invented; we can learn to recognize from visual appearances which surfaces will feel smooth and which rough. Once we are able to move around on our own, we find that, most of the time at least, our senses are giving us reliable information. Trees are where we see them, for example. A little reflection on the implications of Darwin's theory of evolution by natural selection shows us that this is what we should expect. Our senses would not have evolved at all if they did not give us accurate information about the world around us, thus enabling us to avoid possible dangers, such as bumping into trees, falling from the edges of cliffs, or becoming victims to beasts of prey whose approach we had not heard, seen, or possibly smelled. Our senses enable us to live longer, to have more offspring and, therefore, to be "fit" in the Darwinian sense.

Of course, our senses are not completely reliable. We have all experienced optical illusions and, in that strange state between waking and sleeping, heard "sounds" or even "voices" that have no counterpart in the external world. Judgements about colour are notoriously subjective, even if we ignore the complications of colour blindness and jaundice. Most of us quickly learn to cope with these exceptions to the reliability of our sense-data, recognizing subjectivity and illusions for what they are. Some illusions are very persistent, however, and can hold the whole human race in their grip: for example, the illusion that the Earth is stationary and that Sun, Moon and stars revolve around it remained almost unquestioned until about 400 years ago; if opinion polls are to be believed, even many modern people who consider themselves to be educated are still victims of that illusion. All of us have difficulty in recognizing that the apparent impenetrability of solid matter is an artefact of the relative coarseness of our senses, or at least of the macroscopic interactions between material bodies. Most of the time, however, we are right to trust the information that those senses give us about the world in which we find ourselves.

Because our senses are so successful, it is easy to move from trusting what they tell us to assuming that they tell us everything that there is to know about our environment. Many human beings see that assumption simply as common sense, and they can easily forget that it *is* an assumption. Reflect, however, that Darwinian natural selection will lead to the evolution only of senses that help to preserve us from immediate physical danger. There could be aspects of the external world that pose no immediate danger to us so we have not evolved any senses to detect them; just as sight would give earthworms no advantage, so they have no eyes. Our mode of life determines the senses we have and their degrees of acuity; we may well be grateful that we have eyes that see, and equally grateful that we do not share a dog's acute sense of smell, and still more so, that there is no danger that we might, as Alexander Pope put it, "die of a rose in aromatic pain".

It is easy to imagine situations in which we might have different senses and therefore different views of the world. Suppose, for example, like the pit vipers, we had an organ that enabled us to locate warm-blooded animals by the infra-red radiation (or heat) that they emit; or suppose that we could sense magnetic fields directly. Some people do indeed claim to be able to do the latter, but the human race as a whole had to learn about electricity and magnetism by observing their effects on other objects – and in terms of the race's total history, did so only relatively recently. Yet it has been suggested that some migratory birds can navigate by means of the Earth's magnetism, as if they have some kind of built-in "lodestone", and some eels certainly can generate their own electricity, which appears to be used not only as a weapon, but, again, also as a navigational aid in the murky waters in which they live.

If there is life elsewhere in the universe, and not just here on Earth, still more possibilities open up. We may well doubt that any form of life that we know, or can imagine, could appear on a planet orbiting a star that is a powerful emitter of X-rays, although Jill Tarter[1] has drawn attention to many extreme environments here on Earth which we might expect to be hostile but in which living organisms have made a home. As a thought experiment, therefore, let us imagine that intelligent beings could somehow evolve in the vicinity of a stellar X-ray source. The "eyes" of such beings would evolve to be most sensitive to that part of the electromagnetic spectrum which was most intense in the radiation from their "sun" – i.e. the X-rays. The most conspicuous objects in the night-time sky would be other X-ray stars. What we call "normal stars" would be invisible to the astronomers of that planet or, at best, some of the nearest ones, emitting weak X-rays, might appear to them as very "faint" objects. Those astronomers and their physicist colleagues would probably arrive at a very different physics of radiation from ours, which is based on the belief that most radiation is of thermal origin. When our imaginary astronomers and physicists discovered "visible light" (as we had to discover X-rays and radio waves) and learned how to detect it from space, our "normal stars" would seem as exotic to them as X-ray stars seem to us. They would be quite surprised to discover that these extraordinary objects outnumbered those they had previously known by a considerable margin.

All these examples show how our view of the external world is shaped by the sensory equipment that we have. Whatever birds and eels may do, we human beings had to "discover" electricity and magnetism. Static electricity was known in antiquity and was particularly associated with the ability of amber to attract light objects, which is why William Gilbert named such phenomena "electric". Although Gilbert is credited with founding the modern studies of both electricity and magnetism in his book of 1600, *De Magnete*, the properties of the lodestone are now generally acknowledged to have been first observed by the Chinese, and the knowledge had reached Europe by the twelfth century of our era. Current electricity was not known until the late eighteenth century, and only in the nineteenth did Faraday and Clerk Maxwell succeed in incorporating these phenomena into the scheme of physical law. Of course, we can all feel a current, when it passes through our bodies as an electric shock, a sufficiently powerful one being lethal, but we have no sensory equipment that enables us to detect a nearby electric field. I have been within a hundred metres of a lightning strike (perhaps fortunately indoors) and felt nothing – indeed, I did not know until sometime afterwards what I had witnessed. Until the work of Faraday and Clerk Maxwell, electricity was considered mysterious and exciting. Modern dramatizations of Mary Wollstonecraft Shelley's[2] *Frankenstein* show electricity in some form as the means of bringing "the monster" to life. Shelley herself did not write this in her novel because one of the points she wanted to make was Frankenstein's determination that the secret of his creation should die with him, in order to spare the human race the burden that his creation of a living creature had placed on him. In her preface to the second edition, however, Mary Shelley made clear that the novel was conceived while she was with a group of literary people (including Byron and her own husband) when they were discussing the then new experiments on electricity (such as those of Galvani and Volta as well as some more spectacular ones) and that they speculated that electricity (or galvanism, as they called it) might somehow be related to life.

Our modern knowledge of electricity and magnetism is much greater than that gained from the experiments that fuelled the heady speculations of Byron and his friends, yet much of it is less than two-hundred years old. In one sense, Mary Shelley was right about the relation between electricity and life: the workings of the brains and nervous systems of all animals, including ourselves, depend at least in part on tiny electric currents; so, without electricity, we *would* be dead. Yet, throughout most of human history and all of prehistory, we were almost totally ignorant about the causes and nature of electrical phenomena. What else might there be in the external world of which we are still ignorant?

Hackneyed though the quotation has become, Shakespeare[3] had a point when he made Hamlet say to Horatio: "There are more things in heaven and earth than are dreamt of in your philosophy". In the preface to his book *Unweaving the Rainbow*, Richard Dawkins[4] complains about the overuse of that quotation and gives the scientist's reply as "Yes, but we're working on it." The complaint is more justified than the supposed reply; too many scientists turn their backs on the investigation of phenomena that just might show the assumptions of scientific materialism to be untenable. The quotation from J.B.S. Haldane that appears at the head of this chapter and to which I have already referred in the previous chapter makes the same point as Hamlet did and reminds us all that the universe may well be queerer than we *can* suppose.

We have become too used to the phenomena of electricity and magnetism to regard them any longer as "queer"; indeed, as we have seen, we have subsumed them into our scheme of physical law. Mary Shelley's preface is sufficient evidence, however, of how mysterious those phenomena seemed in the early nineteenth century, before the work of Faraday. It would be rash and unscientific to deny the possibility of other phenomena, as yet unrecognized by us, the study of which could lead us to as radically new insights about the universe as did the study of electricity, magnetism, X-rays and radio waves. I claim no originality for the suggestion that our knowledge of the external world depends on the sense-data we receive and on the ways in which our minds interpret them; rather, the idea goes back at least to Kant, who also lived through the time of the early work on electricity. Perhaps there may even be aspects of the universe that cannot be incorporated in the scheme of physical law at all.

Such arguments are, of course, quite contrary to the prevailing modern philosophy of scientific materialism, in which the material universe is conceived of as existing quite independently of us and as being all that there is. Many people are so confident that we have discovered all the types of phenomena there are that they believe we will soon have a "theory of everything". There is no room in this philosophy for paranormal phenomena or religious insight.

Science is limited to sensory knowledge, in this view, or at least to knowledge of phenomena that can be converted into sensory data. Adherents of this point of view would probably argue that the assumption that our senses tell us all that there is to know about the universe is to be preferred because, in the absence of clear evidence, it is the simplest assumption that can be made. They might go on to invoke Karl Popper's criterion that a theory can be termed "scientific" only if it is falsifiable, and to argue that anyone who speaks of the possible existence of aspects of the universe that our senses cannot detect is being "unscientific" – precisely because the suggestion is not falsifiable. We might recall, however, that in the present scientific climate there is serious discussion of the possibility of many "universes" coexisting – a proposal that I will discuss in more detail in Chapter 4 – even though the existence of such other universes is certainly not falsifiable at present, and may never be. The evidence for the existence of realms transcending the physical is, at present, at least as good as the evidence for other physical universes, providing one is open to areas of experience that are not usually considered to be "scientific".

Science has, of course, done very well in interpreting our sensory knowledge and it is perhaps understandable that many have come to assert that our senses suffice to tell us about *all* aspects of the universe. Nevertheless, that assertion is unproven and questioning it is the basis of much of what follows. In order to avoid cumbersome repetition of phrases like "aspects of the world not revealed to us by our senses", I want to choose one word to stand for them. Most of the words that come to mind, "spiritual", "supernatural", "numinous" or "transcendent", for example, have connotations that make me hesitate. For some time, I favoured the word "numinous", which carries suggestions of wonder, or even awe, and has associations with Kant's *noumen*. The adjective was used by Rudolf Otto[5] in his book *The Idea of the Holy*, but he intended it to refer particularly to the non-rational aspects of the Godhead, whereas an important part of my argument is that a rational case can be made for religious belief. I have, therefore, settled on "transcendent" as being the most neutral term available. Only after doing so did a re-reading of *Personal Knowledge* by

Michael Polanyi[6] remind me that he uses the same word in a very similar sense. My argument, then, is that it is quite rational for one to be open to the idea that the world contains entities that transcend our senses.

There are both similarities and differences between this idea of a transcendent realm and Karl Popper's "World 3". In the book, *The Self and its Brain*, which he wrote with John Eccles, Popper[7] spoke of three worlds: "World 1", the world of the senses, "World 2", the inner subjective world, and "World 3", the world of the collected products of the human mind, our culture, things that are not immediately apprehended by the senses and yet which may be said to exist. I am not happy with the division of the world into three, but Popper argues that things in his World 3, even though they are not themselves physical, may be said to be real, because they can affect our behaviour in World 1. As an example of a World 3 object, Popper chose Mozart's *Jupiter* symphony. Of course, he says, it exists in World 1 as the score, or as individual performances, but neither of these *are* the symphony, which exists in World 3, in a way somewhat resembling that in which Plato supposed mathematical ideals to exist. The transcendent realm that I am postulating would certainly contain the things that Popper put in World 3, but I am also open to the possibility that it contains entities that are independent of human creativity – a conclusion to which Peter Dodwell[8] also comes in his book *Brave New Mind*. Indeed, it could be argued that there is a sense in which the *Jupiter* always existed and that Mozart discovered it, rather than created it. A friendly critic has pointed out to me that the same could be said of many very trivial things, but, as will later become evident, I am focussing here on the similarities between artistic creation and scientific discovery, which often seem to share a kind of direct intuition. In Sanskrit, I understand, there is a word, *prathibā*, to denote that kind of intuition of the transcendent without the aid of the senses[9].

I shall discuss in more detail the suggestion of the previous paragraph (which does not necessarily detract from the achievements of Mozart or of any other creative artist) towards the end of the chapter. For the moment, I want to defer further consideration of such matters, and any answer to the arguments of scientific materialism, to ask the question whether, if there really are transcendent (in the sense I am using the term) aspects to the universe, and if some people come by other than sensory means to at least a partial knowledge of them, would those people be greatly honoured and regarded as leaders of thought? A short story by H.G. Wells[10], *The Country of the Blind*, perhaps a modernized version of Plato's allegory of the Cave, is instructive here. The hero of the story, Nunez, arrives by accident in an Andean valley in which the inhabitants have been cut off from the outside world for fifteen generations, as a result of a volcanic eruption. Some time before that, a strange illness caused people living in the valley to go blind, and their children also to be born blind. (Wells displays here what is usually thought to be a Lamarckian view of evolution, rather than a Darwinian one, but he was writing in 1904 and can be forgiven that.) Just before the catastrophe, one of the last of the sighted members of the community had reached the outside world in search of help, so the outsiders had a legend of an isolated tribe in the mountains, and the inhabitants of the valley had legends of a world beyond theirs, but, by the time of the story, neither community took the legends seriously.

When the hero discovers that all the people he encounters are blind, he recalls the proverb "In the country of the blind the one-eyed man is king", and assumes that he will soon lord it over the inhabitants of the valley, whom he supposes to be severely handicapped. He tries to explain the marvels that he can "see" around him, only to find that the very concept of sight is incomprehensible to his listeners, and all words related to light and vision have become meaningless to them. The blind people think that he sounds more like a raving lunatic or, as they put it, "a newly-formed person" than someone of superior knowledge. He tries to explain that he comes from a far country, but the wise men of the tribe have long ago decided that stories of such a place are pure imagination. He discovers, however, that although all around him are blind their other senses have so increased in acuity that his vision gives him only a slight advantage, which is not enough to compensate for their greater numbers, and he is eventually forced to submit to being the slave of one of the landowners. He might have ended his days in that situation, but he falls in love with one of his master's daughters, and she with him. His master is reluctant to consent to the match, however, because of the hero's known peculiarities. At last, Nunez submits to a medical examination. The best local doctors decide that he is potentially sane but subject to delusions. There are two small organs in the front of his face that are unusually sensitive, and the signals they send to his brain cause his delusions. Fortunately, the cure is simple; removal of the organs is a very minor surgical procedure and, if the hero will submit to it, his delusions will cease and he will be able to live a normal life. For the love of the girl, he reluctantly agrees to the operation, but finds at the last minute that he cannot face the loss of sight. He decides that the cliff faces down which he fell into the valley and had thought to be impossible to scale do, after all, offer a means of escape and, understandably if somewhat ungallantly, he leaves his fiancée the night before his surgery is due and returns to the world of the sighted.

Wells was certainly not religious in any conventional sense and probably would have been sceptical of my arguments for the transcendent, but this story of his is a wonderfully pointed parable, and, as many skilled writers do, he leaves the reader uncertain just which side he himself is on. Who is more dogmatic – Nunez, who insists on the superiority of being sighted and resorts to violence when he cannot convince by argument, or the wise men of the valley, who have decided that all talk of a world beyond the one of their immediate experience is meaningless nonsense? There are echoes of Wordsworth's *Intimations of Immortality* in Nunez's insistence that he comes from a far country and in his eventual return to it. There is even an echo of the *Gospel According to St John*[11] and the famous quotation "Greater love hath no man than this, that a man lay down his life for his friends", even though, when it comes to the point, Nunez cannot give up his sight for the woman he loves. Wells' hero actually came off lightly compared with a number of religious and philosophical teachers. Jesus of Nazareth was condemned to a criminal's death, as was Socrates some four centuries earlier; Mohammed felt compelled to flee Mecca for Medina, taking his followers with him. Oriental teachers appear to have fared rather better: the Buddha lived to a ripe old age and died a natural death.

If the transcendent does exist, then, and there are people who have attained in some way a keener awareness of it than most of us have, they may well be regarded as deluded by many of their contemporaries. The most outstanding ones may attract a following and perhaps become more highly esteemed by later generations. It is at least conceivable that the founders of the great religions were people of this sort. In that case, some might object, why are there so many different religions? If the great religious teachers apprehended something that is objectively true about the universe, why did they not all teach the same things? We can perhaps see an answer to this objection if we recall the Darwinian argument that our senses evolved to protect us from immediate physical dangers in the environment, or to give us some sort of advantage in the physical world. Ignorance of the transcendent neither involves that sort of risk nor puts us at any obvious biological disadvantage. Perhaps there are always people with keener-than-average awareness of the transcendent, but since that awareness confers no immediate advantage, the trait is not enhanced from one generation to the next; perhaps it is not even passed on. Just as the earliest creatures to have developed a light-sensitive spot would have only a vague impression of the difference between light and darkness and nothing like the clarity of vision which we and many other creatures enjoy, so people with only a rudimentary sense of the transcendent would apprehend it only very indistinctly.

The Buddha himself told the now well-known story of blind men feeling an elephant. Each one felt a different part of the animal and came up with his own simile for the elephant, according as he had felt tusks, trunk, tail or feet. The Buddha's point in telling the story was to illustrate the difficulty of defining what he called "Buddha nature". Each of the blind men was right as far as his own experience went, but none of them had really understood what an elephant is, because none of them could see the entire creature. We shall not, I think, be departing too far from the spirit of the Buddha's parable if we substitute the word "transcendent" for the phrase "Buddha nature". If we go on to suggest that even the great religious teachers of the human race, although exceptionally aware of the transcendent, were to some extent in the position of the blind men, we have a possible explanation for the existence of so many religions. Many believe that the multiplicity of religions is sufficient evidence that none of them can be true; but it can also be argued that each of the great teachers has seen some aspect of the truth, although none of them grasped it all. That, of course, is unlikely to commend itself to the orthodox of any religion; least of all to orthodox Christians and Moslems, each of whom believes that God's complete and final revelation is enshrined within their own tradition and that other religions are false, or at best incomplete. Another explanation for religious diversity is found in the limitations of all languages that make it impossible fully to describe what is ineffable. All religions use imagery, and I shall comment on that in more detail in Chapter 8. Thus the existence of many religions does not necessarily imply that they are all false, nor is it a bad thing in itself. The diversity of religions becomes bad only when believers in a particular religion claim that theirs, or even just their own version of it, is the only true one. A corrective to that attitude is for all of us to try to learn as much as we can about what other people do believe, and not to rely upon the wilder comments in our media, or the impressions we may have picked up about religious belief and practice without actually examining them. Knowledge brings understanding and respect, even if we cannot share the beliefs of others.

Let us return, however, to consider the situation of those of our contemporaries who might have an awareness of the transcendent only slightly keener than average. We could well imagine some scientist trying to test their talent. Probably they would find it very hard to obtain consistent results – simply because the talent is *ex hypothesi* a weak one. Perhaps some of those claiming to have the talent would commercialize it and, in order to maintain consistency, resort to rather easily exposed trickery, thus discrediting not only their own claims, but those of others. We can easily imagine that sympathetic investigators and sceptical reviewers would soon be engaged in a never-ending battle.

The critics would attack the experimental methods and the statistical analyses of the investigators, who would respond by tightening up their experimental protocols and adopting more rigorous analyses, only to find that their critics were still not satisfied and were demanding more and better evidence.

Precisely the situation just described exists already with respect to so-called *paranormal phenomena.* While I personally think that the evidence for some of these phenomena, telepathy for instance, is strong, many who are sceptical of religion are equally sceptical of the paranormal. Early investigators, such as Rhine and Soal, claimed very high odds against their results being due to chance, as is described by C.D. Broad[12], but the applicability of the rather elementary statistical criteria that they used has been questioned (see e.g. Brown[13]). It is perhaps unsafe, therefore, in our present state of knowledge, to try to argue for the existence of the transcendent, as I have defined it, from that of the paranormal. Nevertheless, there is a certain parallelism between the two: if scientific materialism is correct, neither the transcendent nor the paranormal *can* exist. Our judgement of the credibility of evidence purporting to show that one or the other does exist is heavily influenced by whether or not we believe in scientific materialism, which, in turn, is, or ought to be, affected by the very evidence that we are trying to judge.

The mutual interaction of evidence and belief was expressed in mathematical form by a Unitarian minister, the Revd Thomas Bayes[14] (1710-1761), in a paper published posthumously in the *Philosophical Transactions of the Royal Society*. Bayes tried to express in mathematical (and therefore quantitative) form how we adjust our estimates of the probability of an hypothesis as we obtain new evidence. For a long time this treatment of probability was much less favoured than the later one developed by Laplace (1749-1827). This was partly because Bayes required the assignment of "prior probabilities" to the hypotheses being tested, before any observational evidence was available – and this seemed little better than guesswork. The convinced scientific materialist, for example, will assign a prior probability of zero (implying "definitely not") to the hypothesis "there exist aspects of the universe not detectable by our senses", while the equally convinced religious believer will assign to the same proposition a prior probability of unity (complete certainty). The agnostic will assign a prior probability of 0.5 – implying that the hypothesis and its opposite are equally likely. Obviously, those who have assigned a zero probability to *any* hypothesis are going to look very critically at any evidence that appears to confirm that hypothesis, and will probably find some weakness; they should, however, perhaps apply some of their scepticism to their own assumptions also.

There are several modern formulations of Bayes' theorem which, while different from his own, are equivalent to it. One of them relates the revised estimate of the probability of a hypothesis, given new evidence, not only to the initial assumptions made by the investigator, but to the probability of the evidence found, *given those assumptions*. Thus, the theorem clearly brings out that *the way we assess fresh evidence is dependent on our prior assumptions*. Consider, for example, a group of experiments that appears to provide statistical evidence for the existence of telepathy. To scientific materialists, paranormal phenomena are a priori impossible. Therefore, evidence that appears to support their existence is immediately suspect and *must* be explained away, either as the result of sloppy experimentation or of fraud (including self-delusion). If such people make use of Bayes' theorem, they will have assigned a prior probability of zero to the hypothesis that telepathy is possible and while they cannot deny the experimental results, will probably regard them as having little credibility, believing that they must be at fault in some way. Those sympathetic to the possibility of telepathy, however, will have assigned a high prior probability to the hypothesis that telepathy is possible and look on the new evidence with more favour. This effect that our assumptions can have on how we assess new evidence will be a recurring theme throughout this book. Most of us are honest enough to try, at least, to make allowances for our biases, but the effect that they have on how we assess evidence is subtle, and not always recognized. Of course, if new evidence favouring an initially dubious proposition continues to mount, Bayes' theorem will eventually show that proposition to be increasingly plausible, whatever prior assumptions are made.

In recent decades, Bayes' approach has been found to have many practical applications and has come back into favour. Even as early as 1959, Polanyi, in the book I have already cited, discusses the effect of our prior beliefs on the way we assess probabilities, as a part of his argument that, even in the exact sciences, a *personal* element enters into our knowledge, although he does not mention Bayes' theorem explicitly. Used properly, that theorem can help us to adjust our initial ideas of probability in the light of new evidence. On the other hand, it can be used to give a false sense of precision to estimates of the certainty with which hypotheses such as "the transcendent exists" can be either asserted or denied, as Michael Shermer[15] amusingly showed in an article in *Scientific American.* Perhaps it is a mistake to try to assign *quantitative* probabilities to hypotheses of that sort. What is important is for everyone to state their assumptions clearly, because then, and only then, other people can understand their estimates of prior probabilities, and of the reliability of fresh evidence. We shall encounter several situations to which this caution is relevant. For example, approximately seventy pilgrims to the shrine of Lourdes have experienced healings certified as medically inexplicable by a board of highly qualified medical experts, and these incidents have been officially claimed by the Roman Catholic Church to be miraculous. One of these is discussed in more detail in Chapter 8, with arguments for and against accepting the claim. It cannot be disputed that the healings in question are highly unusual but, of course, anyone may challenge the claim that they were miracles wrought by God. Devout Christians who believe that God does intervene in human affairs will find in each of the Lourdes healings further confirmation of their belief. Scientific materialists, on the other hand, who believe that there is no God, *must* find another explanation – or be converted! Thus, each kind of person assesses the evidence, for which there can be no reasonable doubt, in subtly different ways.

I suggested earlier that Mozart might have discovered his music, which already existed in the transcendent realm, rather than created it. Conventionally, of course, we distinguish between artistic creation and scientific discovery and

I suspect that most artists and most scientists think of themselves as being engaged in very different activities from each other. On the other hand, those whom we regard as the greatest scientists – Newton, Darwin and Einstein, for example – clearly show signs of creativity in their work. Rembrandt was a somewhat older contemporary of Newton and the two men were of vastly different temperaments but if they had been able to meet and talk about their work to each other, they might well have found they had a deep mutual understanding, somewhat as Bertrand Russell[16] and Joseph Conrad did. The idea that music might be discovered rather than created was seriously made a few years ago in a scientific journal by Patricia Gray[17] and others. Indeed, they went further and suggested that *Homo sapiens* was not the first species to "discover" music. The vocalizations of whales and elephants, not to mention the songs of so many different kinds of birds exhibit many of the same characteristics as are found in Western classical music. The idea is no more than speculation at present, but it is thought-provoking.

Mozart himself, at least as quoted by Roger Penrose[18], gave us some grounds for believing that he thought of his work as one of discovery rather than creation, while Leonard Bernstein[19], writing of the slow movement of Beethoven's *Eroica* symphony said that "…it always seems to me to have been previously written in Heaven and then merely dictated to him." Again, Michelangelo is known for speaking of the sculptor's task being to reveal the statue already present in the block of marble. William James[20] took up that idea and suggested that there were perhaps many potential statues in any given block; an individual sculptor would reveal the one most consistent with his or her style and ideas. There is no doubt that any other sculptor confronted with the block from which Michelangelo carved his *David* would have produced a different statue, even if, like Michelangelo, that sculptor had to work within the constraints of cuts already made in the block by another and the subject having been set. One has only to think of sculptors as different in style as Michelangelo and Henry Moore to realize that the former's words cannot be taken literally – yet we have to try to do justice to the conviction apparently felt by at least some of the greatest creative artists that their works were, in some sense, given to them.

Among scientists, Eddington[21] once argued that his contemporary Rutherford had "manufactured" the atomic nucleus, rather than discovered it. He admitted that he was defending an extreme point of view, but his point was precisely to argue that the processes of scientific discovery and artistic creation have more in common than most people suppose. As I hinted earlier, many of the most fundamental discoveries in science are made by intuitive mental processes that resemble artistic creativity and are not simply the result of pure reason applied to the empirical evidence. An intuitive leap is made by the discoverer that is not entirely rational – a point to which I shall return in Chapter 7, when discussing revelation. Much the same is true of mathematics, even though mathematicians talk of discovering their theorems, as is discussed by Peter Dodwell in the book I have already cited; he also writes of the relation between "discovery" and "invention". I prefer Eddington's terminology of "discovery" and "manufacture", since "invention" and "discovery", etymologically speaking, mean the same thing. The Swiss mathematician, Leonhard Euler, a contemporary of Mozart who had a much longer life than did the composer, was one of the most prolific mathematicians who has yet lived. His most famous equation is, perhaps:

$$e^{i\pi} = -1,$$

or, as some people prefer to write it, to include what they regard as the five most significant numbers:

$$e^{i\pi} + 1 = 0.$$

Anyone with enough knowledge of mathematics to know what the symbols e, i and π denote is captivated by the beauty of this equation, quite comparable to the beauty of much of Mozart's music. There is a story that Euler confounded Diderot in a debate at the court of Catherine the Great by quoting to him a mathematical equation and then saying: *"Donc, Dieu existe, Répondez!"* In the oldest version of this story to have been printed (Augustus de Morgan's[22] *Budget of Paradoxes*) the equation is a nonsense one. Perhaps the story got garbled before it was committed to print; if Euler had quoted the above equation, there would have been some point in his statement! Euler and Mozart made very different contributions to human knowledge and experience, yet each of them could have been giving us insights into the same transcendent world.

The beauty of Euler's equation reminds us that even the universe revealed to us by our senses is infused by beauty. I do not mean just the natural beauty of the scenery and vegetation that surrounds us here on Earth, or even the beauty of the starry sky. Since we have learned to explore the greater universe from above the Earth's atmosphere, we have become aware of a weird beauty both in the solar system and beyond. Close-up photographs of the atmosphere of Jupiter, or pictures of star-forming regions taken by the Hubble telescope, have an immediate and wide appeal that suggests that beauty cannot reside entirely in the eye of the beholder. This physical beauty is complemented by the beauty often found in the mathematical equations that describe the physical world. That fact may make us more inclined to think that scientists, mathematicians and artists share some aims and methods, however different their approaches may at first sight seem. Each, in his own or her own way, explores the beauty that surely is in the world in which we live. It does not seem far-fetched to suggest that those whom we call creative (in any of these fields) are sensitive to things in a transcendent realm that most of us cannot perceive by ourselves, but which become perceptible to us by their labours.

References:

[1] Tarter, J., *The Evolution of Life in the Universe: are we alone?* in *Highlights of Astronomy,* **14**, 14-29, 2007, Cambridge University Press.

[2] Shelley, M. Wollstonecraft, *Frankenstein or the Modern Prometheus*, Author's Introduction to the 2nd Edition, 1837, reprinted and edited by S.J. Wolfson, together with text of the 1817 1st Edition (Longman, New York, 2003), pp. 186-191.

[3] Shakespeare, W., 1600, *Hamlet* Act 1 Scene 5.

[4] Dawkins, R., *Unweaving the Rainbow: Science, Delusion and the Appetite for Wonder* (Houghton Mifflin & Co., Boston & New York, 2000), p. xi.

[5] Otto, R., *The Idea of the Holy* (English translation by J.W. Harvey, 1923 and 1950 of *Das Heilige*, Oxford University Press, 1917).

[6] Polanyi, M., *Personal Knowledge* (University of Chicago Press, 1959; corrected edn. 1962).

[7] Popper, K.R., and Eccles, J.C., *The Self and its Brain* (Springer Verlag, Berlin, 1977), Chap. P2, and pp. 449-50.

[8] Dodwell, P.C., *Brave New Mind* (Oxford University Press, 2000), pp. 63-4, 191-2.

[9] I am indebted to Harold Coward for pointing this out to me.

[10] Wells, H.G., *The Country of the Blind*, *Strand Magazine*, April 1904, reprinted in several collections, e.g. H.G. Wells*, Selected Short Stories* (Penguin Books, Harmondsworth, 1958).

[11] *Gospel According to St John*, Chap. 15, v. 13.

[12] Broad, C.D., *Lectures in Psychical Research* (Routledge and Kegan Paul, London, 1962).

[13] Brown, G.S., *Nature*, **172**, 154-156, 1953.

[14] Bayes, T., *Phil. Trans. Roy. Soc.*, 1763, pp. 376-418.

[15] Shermer, M., *Scientific American*, **291**, No. 1, *God's Number is Up*, p. 46, 2004.

[16] Russell, B., *The Autobiography of Bertrand Russell, Vol. 1* (George Allen and Unwin Ltd, London, 1961), p. 341.

[17] Gray, P.M., Krause, B., Atema, J., Payne, R., Krumhansi, C. and Baptista, L., *Science*, **291**, 52-54, 2001.

[18] Penrose, R., *The Emperor's New Mind* (Oxford University Press, 1989), p. 423. Penrose quotes an English translation of Mozart's letter given by Jacques Hadamard in *The Psychology of Invention in the Mathematical Field* (Princeton University Press, 1945), p. 16. Hadamard himself cites another two works, not available to me. Since neither the date nor the recipient of the letter is given, I have been unable to find the original in Mozart's published correspondence.

[19] Bernstein, L., *The Joy of Music* (Simon and Schuster, New York, 1959), p. 29.

[20] James, W., *Principles of Psychology*, Volume 1 (Henry Holt & Co., 1890; Dover Reprint, 1950) Chap. IX, p. 288.

[21] Eddington, A.S., *The Philosophy of Physical Science* (Cambridge University Press, 1939), Chap. VII.

[22] de Morgan, A., *Budget of Paradoxes,* 1st edn. 1872, 2nd edn. 1915 (Dover Reprint with an introduction by E. Nagel, New York, 1954), Vol. II, pp. 1-4.

CHAPTER 3
Belief in God

L'amor che move il sole e l'altre stelle.

Dante Alighieri, *La Divina Commedia*

Dante, as we have seen, was thoroughly versed in the Ptolemaic astronomy of his day, and the last word of each of the three main sections of his great poem was, in Italian, *stelle*, or stars; yet he found no difficulty in writing of God as the love that moves the Sun and the other stars. He thought, of course, in Aristotelian terms, regarding the planets as "movable stars" (as opposed to the fixed stars) and supposing that bodies needed to be subjected continuously to a force in order to remain in motion; but to most modern minds, that last line of *The Divine Comedy* can be no more than a poetic conceit. We do not think of love as a force in the sense that word has been used in physics since the time of Newton. Material objects, it is argued, can be moved only by the forces recognized in modern physics, and in the philosophy of scientific materialism those are the only forces that there are.

Scientific materialists have thought of a variety of names to describe themselves: atheist, agnostic, humanist, sceptic, rationalist and, most recently, "bright". Of all these names, "rationalist" seems to me to be the least fair; using it cleverly suggests that those who hold other opinions are in some sense "irrational". I have tried to argue in the previous chapter that it is not irrational to believe in what I have called "the transcendent". I rather deliberately left God out of the picture, but now I propose to extend my argument to urge that neither is belief in God necessarily irrational. Dante would no doubt have approved of this extension, but it *is* an extra step. While those who are not convinced of the existence of the transcendent will certainly not go on to consider the existence of God, it is possible for people to accept the former without being theists. Buddhists provide a good example of that attitude, and a Western writer who appears to have a similar position is the philosopher Sam Harris[1]. In his recent book, *The End of Faith*, he shows himself to be quite open to the possibility of our personalities surviving in some way the deaths of our bodies, and somewhat sceptical of the opinion that our consciousnesses can be reduced to the activity of neurons in our brains. Yet he is strongly critical of theistic belief, especially of that embodied in the three Abrahamic religions of Judaism, Christianity and Islam (to list them in their historic order).

Many people are put off by the word "God" because of the connotations that have grown up around it. The Second Commandment, which Jews and Moslems have kept much more strictly than Christians have, forbade any representation of God, with good reason. Not only have Christians permitted portrayals of Jesus, to Whom they have attributed Divine status, but they have also permitted portrayals of God the Father as an old man with a long white beard, sitting on a throne in the sky. To that extent, others can be forgiven for supposing that that is how Christians envisage God. I shall discuss in Chapter 8 the role of picture language in both religion and science, and in Chapter 10 what it may mean to believe in a personal God. For the moment, I merely wish to stress that I, at least, do not think of God as having a human form, still less as being male. To emphasize this last point, I shall endeavour to avoid the use of a pronoun for God; in the few cases in which I cannot do so without great linguistic clumsiness, I shall fall back on the conventional "He". I hope that those readers for whom this is an important issue will recognize this practice for the linguistic convenience that it is, and not assume that it is meant to carry any implications about the nature of God, other than that the impersonal "It" would be an inappropriate pronoun to use.

To return to the question of whether or not a rational case can be made for the existence of God, that is, presumably, exactly what St Thomas Aquinas[2,3] (who was one of Dante's sources of inspiration) was trying to do in his famous "five ways*". Indeed, attempts to prove the existence of God have for a long time been considered part of what is known as "natural theology" based on the belief that all human beings have access to some knowledge of God, regardless of what religious beliefs they hold, or even if they hold none at all. The essence of natural theology is summed up in an argument found in the book we know as the *Wisdom of Solomon*[4] (and echoed in the early chapters of St Paul's *Epistle to the Romans*): "…through the grandeur and beauty of the creatures we may, by analogy, contemplate their author". Such ideas fell out of favour among theologians themselves during much of the twentieth century, at least partly because of the excesses to which they were sometimes taken in the nineteenth, as we shall see in the next chapter, but a justification for them was published some years ago by James Barr[5], in his book *Biblical Faith and Natural Theology*.

So much has been written about the five ways* and other purported proofs of God's existence (especially, of course, St Anselm's ontological argument) that I doubt if even those who have made a lifetime study of the subject can have read all of the resulting literature. Perhaps it is rash, therefore, for a rank amateur in both philosophy and theology to express any opinion at all about those arguments. It is significant, however, that the arguments have stimulated so much debate, and are still quite often cited by those who wish to discredit any form of theistic belief. If they were the conclusive "proofs" that they have sometimes been represented to be, then, surely, by now, everyone who had studied them would be convinced by them. Equally, if they were as completely fallacious as some scientific materialists make them seem, everyone would by now have seen through them. That the arguments are still discussed is evidence that they have some persuasive power, even if they fall short of being complete proofs in the sense that most of us now use that term.

---

* The first three ways purport to show, respectively, that there must be a *prime mover*, a *first cause*, and a *necessary being*. The other two ways are described later in the text.

Our notions of proof are conditioned by the way the word "proof" is used in Euclidean geometry: a theorem is stated and shown to be true by an argument formulated as a series of logical steps, at the end of which we triumphantly write the letters "QED". At first sight, this looks like a genuine way of increasing the number of things we can be certain about. Even Einstein, as a young boy, thought that Euclidean geometry was a way of obtaining certainty about the external world, although, in later life, he took the lead in introducing the notion that other kinds of geometry might, after all, describe that world better. Although the notion of non-Euclidean geometries, developed in the nineteenth century, still seems strange to non-mathematicians, some geometries of that kind are easy to construct. Euclid's theorems about triangles, for example, do not apply to triangles formed on the surface of a sphere. They may seem to apply to triangles drawn on the surface of the Earth, if those triangles are so small that we may neglect the fact that the Earth's surface is curved, but we cannot neglect that curvature when we survey and map whole continents. Consider Euclid's theorem that the angles of a triangle add up to two right angles, or 180°; it applies only to triangles drawn on a plane surface. A spherical triangle may be formed on the surface of the Earth by taking any two meridians of longitude that are 90° apart and the arc of the equator between them. It is immediately obvious that the sum of the angles of this triangle must be greater than two right angles, since each of its three angles is 90° and their sum must be 270°, or *three* right angles. That is an extreme example, but, in general, the sum of the three angles of any triangle drawn on the surface of a sphere will be greater than two right angles.

This digression serves to make clear that Euclidean geometry, and logical argument in general, do not necessarily lead to new knowledge. The "proof" that the angles of a triangle add up to 180° simply makes explicit what is already implicit in Euclid's axioms – which define what he meant by such terms as "point", "line", "plane" and "parallel". If, however, Euclid's axioms do not describe the true nature of the space in which our universe is situated, then they do not apply and, even though there is no logical flaw in the argument, the theorem is no longer true. (I do not wish to argue that the whole of mathematics is tautological, although that view was held by several distinguished mathematicians about a hundred years ago; see the discussion at the end of the previous chapter, about discovery and manufacture.)

I doubt if Aquinas thought of his five ways as equivalent to mathematical proofs. According to F.C. Copleston[6], as he wrote in his book *Aquinas*, St Thomas believed that metaphysical reflection on the empirical facts on which he based his arguments (*some things move*, for example, or *some things seem to change*, or *inanimate things appear to work together for an end*) would lead a fair-minded person to the conclusions that Aquinas himself found. If we persist, however, in seeing the five ways as analogous to geometrical proofs, we have to admit that, even if there were complete agreement that there are no logical flaws in the arguments themselves, which there is not, those who dislike the conclusion can still contest the axioms. Aquinas did not state his axioms quite as explicitly as Euclid did, but there is one assumption that is common to each of the first three proofs – that we cannot have an infinite regress of cause and effect. Strictly speaking, this is not an assumption, since Aquinas goes to some lengths to try to prove that we cannot have such a regress. Kenny[7], however, discusses that argument at length and concludes that it fails. Therefore the proposition that there cannot be an infinite regress is effectively an assumption for us. If that assumption is granted, there must be an unmoved mover, an uncaused cause, and a "necessary being" as the source of all contingent beings such as ourselves. There is a sense in which these three arguments are really only one. (It is not important that Aquinas, basing his arguments on Aristotelian physics, did not know Newton's concept of inertial motion – a body continuing in a state of rest or uniform motion in a straight line unless acted on by a force. We can easily substitute "acceleration" for "motion" in the first proof and end up with a "prime accelerator". Moreover, as Kenny stresses, "motion" had a somewhat broader meaning for Aquinas than the word does for us: he thought of growth and some other forms of change as being a kind of motion.)

The important point is that the assumption about the infinite regress is open to challenge. In his discussion of the five ways, Copleston argues that Aquinas was not thinking of an infinite regress of causes in *time*, but of a hierarchy of causes, all acting on each other here and now. Therefore, he says, in a dart that I suspect was aimed at Bertrand Russell, arguments about infinite mathematical series are irrelevant to the question of whether or not the five ways are valid proofs. That may be so, but I think most of us find it equally hard to imagine *either* an infinite regress of cause and effect *or* an uncaused cause (although some atomic and sub-atomic phenomena such as radioactive decay seem to be uncaused in the sense that we cannot predict when any *individual* atom will decay). Astronomers, regardless of whether or not their expertise is in cosmology, are often asked whether the universe is finite or infinite. The questioner usually finds it just as difficult to imagine either infinite space, or finite space with nothing (in the sense of "no thing") outside, difficulties that Kant fully recognized in posing his antinomies. For modern minds, the first three of the five ways would still present that sort of difficulty, even if they were otherwise above criticism. Interestingly enough, according to the Dalai Lama[8], some Buddhist thinkers have used the fact that a creator God must stand outside the chain of cause and effect as an argument *against* the existence of such a being.

Any intelligent child told that God made everything asks the inevitable question "Who made God?", only to be told that God just "is". That, perhaps, is a way adapted to the understanding of a child of expressing Aquinas's concept of a necessary being – a concept that has been the target of many criticisms of Aquinas's third way, at least since Kant argued that the phrase is a misuse of language. Thus Kant disposed simultaneously of Aquinas's third way and of St Anselm's ontological argument – a conclusion that I believe most modern philosophers accept, although Kenny[9] is an exception. I shall not discuss the ontological argument any further, except to say that I rather like the comment Bertrand Russell[10] once made about it: "…it is easier to feel convinced that [the argument] must be fallacious than it is to find out precisely where the fallacy lies." Nevertheless, many of us feel the need to search for an explanation of why there should be anything at all. Kant's argument that we cannot meaningfully talk of a "necessary being" may be intellectually convincing, but the more modern suggestion by E.P. Tryon[11] that "our Universe is simply one of those things that happen from time to time" scarcely seems satisfactory – and, as Peacocke[12] has pointed out, the quantum vacuum in which things happen from time to time is itself something existent in need of an explanation.

Aquinas's fourth way is probably the least convincing of the five to modern minds. Unlike the first three, which are clearly Aristotelian, it is more of a Platonic argument. Since beings can have different degrees of various qualities, the argument runs, there must be a being that is perfect in all respects. It does not seem necessary to discuss this way any further. The fifth way is a version of the argument from design, which deserves a full-length discussion of its own and which I postpone to the next chapter.

Aquinas's five ways, of course, with the possible exception of the fourth, do not tell us much about the nature of God. The qualities traditionally associated with the Christian God – love, mercy, omnipotence, omniscience, etc. – are not necessarily to be found in a prime mover, a first cause or even a necessary being. Naturally, Aquinas recognized that himself and, therefore, relied on revelation as well as reason. His aim was to show that, even without revelation, some idea of God could be reached by reason, and, for him, Aristotle was the supreme exponent of unaided human reason. It is not irrational, according to Aquinas, to believe in God. Some might contest even this modest claim, arguing that it *is* irrational to believe in someone (or something) whose existence we cannot prove. Once again, an example from mathematics, Gödel's celebrated incompleteness theorem, is illuminating. All mathematicians believe in the theorem, which states that in any logical system we will encounter some propositions that can be neither proven nor disproved within the axioms of the system. The principle of (the rather lengthy and complicated) proof is to show that the theorem itself is such a proposition – so we believe in the theorem *because* we cannot prove it. On a more practical level, we do, every day, act on the basis of probabilities, rather than certainties. Indeed, there are many situations in which we would consider it irrational *not* to act on the basis of the best information that we have, even though that information may be uncertain, and the consequences of acting on it even more so.

The real question with Aquinas's five ways, therefore, is whether or not they carry enough conviction to be the basis for our action. The answer to that will obviously vary from person to person. Those already inclined to believe in God will see in at least some of the five ways a degree of rational support for their belief. Possibly, some people, uncertain whether to believe or not, will be won over by one or other of the arguments, but thoroughgoing sceptics will remain unconvinced. Of course, Aquinas and Anselm are not the only people who have tried to prove the existence of God; Leibniz, in particular, produced arguments of his own, some of which were modifications of earlier arguments. Even Bertrand Russell conceded that Leibniz's arguments are harder to refute than are those of Aquinas – which may be one reason why scientists turned amateur philosophers tend to leave them alone. The five ways and the ontological argument have, however, become the best known outside the circle of professional philosophers, and are therefore more widely discussed. If we claim that the five ways, or any of the other arguments purporting to prove the existence of God, are conclusive, we invite those who do not believe to look for flaws in the arguments – and they have made a pretty good job of finding them. If, on the other hand, we assert that the five ways have sufficient plausibility for us to base our lives on the assumption that the conclusion to which they point is true, others may disagree with our estimate of plausibility and say that our belief in God is based on faith rather than knowledge, but they cannot legitimately accuse us of irrationality.

Faith has a bad name amongst sceptics, who often refer to it as "blind faith" and see it as belief in propositions for which there is either no evidence at all, or only evidence that will not withstand modern critical analysis. Among contemporary writers, Richard Dawkins[13] (*The God Delusion*) and Sam Harris (in the book already cited) attack this sort of faith, which they see as the source of many of the evils of the modern world and in history. Faith, however, is not simply a matter of believing *propositions* and is much more analogous to trust in a *person*. Granted, to trust someone you must first believe that he or she exists and is trustworthy; nevertheless the Greek word that is usually rendered as "faith" in English versions of the New Testament could equally well be translated as "trust". (Harris knows of that meaning, since he mentions it in his book, p. 64, but he continues to attack his own definition of "faith".)

It may seem difficult, given the present condition of the world, to trust in a God who is both all-powerful and loving, but it is not necessarily irrational to continue trusting despite appearances, and that is not the same thing as believing a proposition without evidence. I trust my closest friends because I have known them for a long time. I have some idea, at least, of how their minds work and of their motives. I can even sometimes predict how they will react to new ideas or unusual situations. If I hear a rumour that one of those friends has done something dishonourable, I am likely to dismiss it because I *know* the person – but I can produce no argument that will convince those who do not know the person as I do. Up to this point, most people would act similarly in such a situation and I think that everyone would commend my loyalty to my friend, even if they did not share it. Suppose, however, that the rumours increase, or my friend is even charged with a criminal offence. If I am worth my salt, unless my friend confesses, I ought to maintain his or her innocence until a guilty verdict is reached. Even then, I ought to look for extenuating circumstances and, even more importantly, help that friend, in due course, to be reintegrated into society. In extreme cases, since miscarriages of justice do occur, I ought to be prepared to fight for a reconsideration of the verdict. That is what having faith in a friend means: it is not entirely rational and yet I can give good reasons for my loyalty. Belief or trust in God should be something like that and is certainly more than giving intellectual assent to the proposition "God exists".

For many people, the analogy of God as a friend to be trusted resonates more clearly than the notion of God as a hypothesis to be proved, or at least demonstrated. Such people are apt to be impatient with theological arguments, because their own experience of life has convinced them of the reality of God. Sir Arthur Eddington[14], the great Quaker astronomer, expressed this idea well:

> Theological or anti-theological argument to prove or disprove the existence of a deity seems to me to occupy itself largely with skating among the difficulties caused by our making a fetish of this word [i.e. existence]. It is all so irrelevant to the assurance for which we hunger. In the case of our human friends we take their existence for granted, not caring whether it is proven or not. Our relationship is such that we could read philosophical arguments designed to prove the non-existence of each other, and perhaps even be convinced by them – and then laugh together over so odd a conclusion. I think it is something of the same kind of security that we should seek in our relationship with God. The most flawless proof of the existence of God is no substitute for it; and if we have that relationship the most convincing disproof is turned harmlessly aside. If I may say it with reverence, the soul and God laugh together over so odd a conclusion.

The twentieth-century astronomer was, perhaps, not far removed in spirit from Dante, even though their concepts of the cosmos differed vastly. But Dante, influenced by Aquinas, would probably have placed more value than Eddington did on theological argument. On the other hand, Aquinas himself, towards the end of his life, had a religious experience so profound that he dismissed all that he had previously written as so much straw, and ceased to write any more. The kind of personal conviction that Eddington had throughout his adult life, at least, and Aquinas apparently acquired towards the end of his life, does not require buttressing by theological argument, any more than we ourselves need to refute the arguments of a solipsist to convince ourselves that our friends exist. My chief concern here, however, is to stress that there is a rational basis for "faith". Personal convictions, as we have seen in the analogy of human friendship, cannot easily be conveyed from one person to another, while rational argument, even if it falls short of "proof" can be used to influence others. Intellectual assent is, after all, a necessary part of the whole relationship; we need both the personal conviction and the intellectual argument, so far as the latter goes.

In *The God Delusion*, Richard Dawkins dismisses even this approach to belief in God, comparing it to the belief that some children have in imaginary playmates who seem very real to them but who are supposed by the rest of us to be entirely the inventions of the children concerned. He even quotes an entire poem by A.A. Milne (*Binker*) about such an imaginary companion. I do not recall that either I or any of my contemporaries had such an imaginary companion, although it is apparently common for a child to forget the companion completely once the need for him or her has been outgrown[15]. The phenomenon is well established, however, and is discussed by John Geiger[16] in his recent book *The Third Man Factor*. Geiger suggests that such imaginary playmates have something in common with the "presence" sensed by many people in extreme danger, which often guides them to safety. I shall discuss that phenomenon again in Chapter 6. Most children who have such companions quickly grow out of the need for them but, if Geiger is right, the companions are something more than make-believe or hallucination, being a creation of the brain of a child, or a person in danger, to fulfil a real need. No doubt, some people would put God in the same category, but, even if that is justified, it *is* a category of something more than make-believe. Even the child's imaginary playmate is not quite delusion and the "Third Man", whatever it may be, has helped many people to survive incredible ordeals. The considered opinion of a mature and highly intelligent adult, such as Eddington, deserves at least a similar respect.

As already mentioned, when we proceed beyond arguing for the existence of God and add claims, such as are found in Judaism, Christianity and Islam, that God is all-powerful, all-good, loving and compassionate, we run into new difficulties. Non-believers can reasonably point to a great deal of counter-evidence to these supposed properties of God, most pointedly in the evils let loose on the world by the more fanatical believers of all religions. The problem of evil is, of course, an age-old one and another about which much has been written. I will explore it in a little more depth later (in Chapter 10), but at this point I wish to focus only on the question of the existence of God. Those who already accept the existence of the transcendent will, on the whole, be inclined to weigh the evidence differently from those who do not (another example of Bayes' theorem in probability), even if, like Sam Harris and the Buddhists, they do not go on to accept the existence of God.

Unfortunately, the prominence given by much of the Christian Church to the recital of creeds as an act of worship strengthens the impression that religious faith is primarily a matter of assenting to propositions – and to propositions that, if not entirely incredible, certainly run counter to normal human experience. Reciting a creed at least appears to involve the giving of assent to some very definite statements about Divine activity in the world which present real problems to a modern scientific mind. I grew up, and have remained, within a branch of the Church which maintains the creedal tradition, and must have recited both the Apostles' and the Nicene Creeds more times than I care even to try to count, but I admit to my own mental reservations and exercise a very broad freedom of interpretation. Not all Christians feel the need to recite creeds; the Quakers get on very well without them. Once again, we turn to Eddington[17], who was quite critical of the practice of reciting creeds as an act of worship:

> You will understand the true spirit neither of science nor religion unless seeking is placed in the forefront.
>
> Religious creeds are a great obstacle to any full sympathy between the outlook of the scientist and the outlook which religion is so often supposed to require. I recognize that the practice of a religious community cannot be regulated solely in the interests of its scientifically-minded members and therefore I would not go so far as to argue that no kind of defence of creeds is possible. But I think it may be said that Quakerism in dispensing with creeds holds out a hand to the scientist… The spirit of seeking which animates us refuses to regard any kind of creed as its goal.

And a little later he continued:

> Rejection of creed is not inconsistent with being possessed by a living belief. We have no creed in science, but we are not lukewarm in our beliefs. The belief is not that all the knowledge of the universe that we hold so enthusiastically will survive in the letter; but a sureness that we are on the right road. If our so-called facts are changing shadows, they are shadows cast by the light of constant truth. So too in religion we are repelled by that confident theological doctrine which has settled for all generations just how the spiritual world is worked; but we need not turn aside from the measure of light that comes into our experience showing us a Way through the unseen world.
>
> Religion for the conscientious seeker after truth is not all a matter of doubt and self-questionings. There is a kind of sureness that is very different from cock-sureness.

Although some people have thought that, at times, Eddington slipped into cock-sureness, his distinction is valid. Faith, trust or belief (whichever word you prefer) in God should be the sureness that the transcendent is real and that we are on the right road, not the cock-sureness that we know God's will and are doing it in our lives. We do not have to look very far or long in today's world to see the evil that is often perpetrated by that kind of cock-sureness.

References:

[1] Harris, S., *The End of Faith*, 2nd edn. (W.W. Norton & Company, New York and London, 2005).

2. Aquinas, T., *Summa Theologiae*, 1274, Latin text with English Translation (reprinted by Eyre and Spottiswoode, London, and McGraw-Hill, New York, 1963). There are also parallel passages in the *Summa contra Gentiles*.
3. Kenny, A., *The Five Ways* (Routledge and Kegan Paul, London, 1969) also gives Latin text and English translation, together with a detailed discussion.
4. *The Wisdom of Solomon*, Chap. 13, v. 5. The translation given is taken from the *Jerusalem Bible*.
5. Barr, J., *Biblical Faith and Natural Theology* (Clarendon Press, Oxford, 1993).
6. Copleston, F.C., *Aquinas* (Penguin Books, Harmondsworth, 1955).
7. Kenny, A., see reference 3, pp. 23-27.
8. Tenzin Gyatso (14th Dalai Lama), *The Universe in a Single Atom* (Broadway Books (Random House), New York, 2005), p. 84.
9. Kenny, A., see reference 3, p. 2.
10. Russell, B., *History of Western Philosophy* (George Allen and Unwin Ltd, London, 1946), p. 568.
11. Tryon, E.P., *Nature*, **246**, 396-397, 1973.
12. Peacocke, A.R., *Theology for a Scientific Age*, enlarged edn. (SCM Press, London, 1993), p. 101.
13. Dawkins, R., *The God Delusion* (Houghton Mifflin Company, Boston and New York, 2006), pp. 347-9.
14. Eddington, A.S., *Science and the Unseen World* (George Allen and Unwin Ltd, London, 1929), p. 43.
15. Gopnik, A., *The Philosophical Baby: What Children's Minds tell us about Truth, Love and the Meaning of Life* (Farrar, Straus and Giroux, New York, 2009), p.52.
16. J. Geiger, *The Third Man Factor* (Penguin Group (Canada), Toronto, 2009).
17. Eddington, A.S., see reference 14, pp. 55-56.

CHAPTER 4
The Argument from Design

> Many worlds might have been botched and bungled, throughout an eternity, ere this system was struck out.
>
> David Hume, *Dialogues Concerning Natural Religion*

The first three of Aquinas's five ways, as we have seen, are Aristotelian arguments and, even if there were no logical flaws in them, all that they could prove would be the existence of an Aristotelian god: impersonal, aloof from the world, unchanging, and unconcerned about the fate of individual human beings. In adapting the arguments to Christian apologetics, Aquinas was well aware that additional input was needed to establish the existence of any being at all like the Christian God. For Aquinas, of course, that input came from revelation through the Scriptures, as interpreted by the Church, but that is a topic we shall leave for Chapter 7. The fourth way, as we have also seen, is more Platonic than Aristotelian and introduces perfection as an attribute of the Godhead. The fifth way introduces us to God as designer, and is possibly the oldest of all purported proofs of the existence of God. Plato knew the argument and rejected it, so it had been advanced even before his time. The opening verse of the nineteenth psalm, "The heavens declare the glory of God and the firmament sheweth his handywork", can be seen as a poetic enunciation of the argument and, whoever wrote them, the Psalms are certainly older than Plato.

The argument from design is not only very old, it is also very seductive; even some of its critics have dismissed it only with reluctance. Hume[1], in his

Dialogues Concerning Natural Religion, presents two characters, Cleanthes and Philo, who respectively argue for and against a designed universe. Hume's use of the dialogue form enables him to keep his own views in the background, but it would clearly be consistent with his other thinking to suppose that he rejected the argument himself. Kant, too, treated the design argument with respect. Perhaps one reason why some modern evolutionary biologists react so strongly against any suggestion of design in evolution is that they, too, have felt the seduction of the argument from design, even as they reject it intellectually. The argument has had different histories in the physical sciences (mainly in astronomical contexts) and the biological, and, in our own time, has resurfaced within each of these branches of science, in the first as the so-called "anthropic argument" and in the second as "Intelligent Design". It is worth tracing something of these two distinct histories.

Among modern scientists, Newton was one of the first to use the argument from design. He was impressed by the fact that all the planets and satellites known to him moved around the Sun in the same sense and in orbits that were very nearly in the same plane. He believed that this could be explained only by direct Divine intervention at the time that the solar system was created. God, according to Newton, put the planets where He wanted them, and gave them the velocities that would keep them within the range of distances from the Sun that He also wanted. Only the comets moved far out of the plane or revolved in different senses, and they again according to Newton had a role to play in the rejuvenation of the Sun and so they, too, were Divinely ordained. The English translation of Newton's own Latin words could almost have been written by a modern exponent of Intelligent Design: "This most beautiful system of the sun, planets, and comets, could only proceed from the counsel and dominion of an intelligent and powerful being."[2]

This argument was not challenged, on scientific grounds at least, until near the end of the eighteenth century when the French astronomer Pierre Simon, Marquis de Laplace[3] (1749-1827), sometimes called the French Newton, gave a mathematical treatment of the so-called "nebular hypothesis" of the origin of the solar system that had earlier been proposed in qualitative terms by Immanuel Kant[4]. In Laplace's theory, the motion of the planets in very nearly the same plane, and in the same sense around the Sun, were natural consequences of the way in which the solar system had been formed and he had no need to invoke God to explain these characteristics. That is the point of the famous anecdote about the conversation between Napoleon and Laplace, in which Napoleon asked what the role of God was in the creation of the solar system and received the reply: "Sire, I have no need of that hypothesis." (Although, as far as I can trace, no contemporary record was made of Laplace's actual words, we do know that there was a meeting between Napoleon and Laplace, at which the famous English astronomer, Sir William Herschel, was present. Herschel kept a diary of his visit to France and wrote in it that the origin of the solar system, and the role of God in it, had been discussed by Laplace and Napoleon[5]. I sometimes suspect that the oft-quoted phrase is what Laplace wished he had said after he got home!) Although Laplace was no friend of organized religion and seems to have been an atheist, his intention in making this reply, if he did, was probably not so much to proclaim his atheism as to make clear that, while even Newton could conceive of no mechanism except Divine intervention to explain the observed properties of the solar system, the French Newton had no need of that hypothesis. Newton himself had fallen into the fallacy that we now call the "God of the gaps", invoking God to explain what the science of his day could not. This is always a danger for those who would argue from design, since subsequent scientific developments may reveal a natural explanation, as Laplace showed in this instance.

Although a modernized version of the Kant-Laplace hypothesis is again the most favoured theory of the origin of the solar system, in the late nineteenth century and early twentieth it fell temporarily out of favour for technical reasons that need not trouble us here. In its place arose a tidal theory associated with the names of T.C. Chamberlain and F.R. Moulton, and much popularized by Sir James Jeans, in which it was supposed that a passing star pulled a massive filament of matter out of the Sun, from which the planets and other bodies in the solar system eventually formed. In this theory, too, the facts that the planetary orbits are all nearly in the same plane and that the planets all revolve around the Sun in the same sense are natural consequences of the way in which the planets are supposed to have been formed. Indeed, no theory that did not explain these facts as quite natural consequences of the mechanism of formation would be for long entertained as even a plausible one by modern astronomers; so far have we come from Newton's conviction that Divine Providence was the only possible explanation of those facts!

Of course, it is still possible to argue that God laid down the laws that ensure the co-planarity of the orbits and the sense of revolution of the planets within those orbits, and that He *is*, therefore, responsible for the "design" of the solar system. As one well-known hymn puts it:

> Praise the Lord, for He hath spoken;
> Worlds His mighty voice obeyed;
> Laws which never shall be broken
> For their guidance hath He made.

The precise date of this hymn (a paraphrase of Psalm 148) is not known; it is found in a collection[6] published in 1796, the very year in which Laplace published a preliminary account of his theory, and it reflects something of both Laplace's famous determinism and eighteenth-century deism, for all that modern Christians still sing it heartily. Indeed, in astronomical contexts, the argument from design resembles the first three of Aquinas's arguments in that, if it proves anything, it proves only the existence of a deist god, who created the universe, laid down the laws by which that universe should unfold, and set everything in motion. Such a god might well remain indifferent alike to the fates of the universe itself and of the sentient creatures, including human beings, who actually live in it, and would resemble Aristotle's First Cause, rather than the Christian God Who is said to be a God of love. This history of the theories of the origin of the solar system, therefore, illustrates two dangers of the argument from design: the designer god may be rendered unnecessary by new advances in science, and may not have all the properties ascribed to God, either in Christianity or in other religions. Indeed, as our consideration of the biological forms of the argument will show, a designer god may even have some properties incompatible with those ascribed to God by Jews, Christians and Moslems alike.

As we have seen, by the time that modern theories of the origin of the solar system had disposed of Newton's notion that only Divine Providence could explain some of its properties, astronomers were beginning to explore the vastness of the universe beyond that system. Design is not so immediately obvious there. Copernicus upset some aspects of the perceived design when he displaced the Earth from the centre of the universe and reduced it to one of the less significant members of the solar system. Then, early in the twentieth century, the Sun was found not to be at the centre of the Galaxy; Hubble and others are often portrayed as having continued and extended the process, when they showed that even the Galaxy was only one of many that make up the universe. Shortly after that, it became clear that the central position of our Galaxy is only apparent. Any observers there might be in other galaxies (or "island universes" as they used to be called) would have exactly the same impression that we have: all other galaxies would appear to be receding from *them* with speeds proportional to their distances. Meanwhile Darwin had told us that we were, biologically speaking, not essentially different from the rest of the animal creation, and Sigmund Freud, whose theories were much more influential in the twentieth century than they appear to be now, represented our cravings for purpose in the universe and for God simply as responses to our subconscious needs. It became fashionable to talk of the "Copernican principle", or the "principle of mediocrity" – an assertion that the human race is not privileged in any way and has no special position in the universe.

Partly in reaction to that assertion, Brandon Carter[7] pointed out that, although we do not observe the universe from any privileged position (such as its centre, if there were such a place) yet beings like us can exist only in a universe with some very special properties and even in such a universe only at special locations in space and time. As he put it: "…what we can expect to observe must be restricted by the conditions necessary for our presence as observers." Expressed like that, the idea is uncontroversial and may even seem tautological – although it can be used to make predictions. The idea has become known as the "anthropic principle" and has been expressed in various ways; some have even argued that the universe *necessarily* has the properties required for the appearance of life – a much more questionable proposition. Attention has focussed on the four basic forces that we encounter in the universe: gravity, electromagnetic force, and the weak and strong nuclear forces. The relative strengths of these forces determine the properties of matter. Gravity is by far the weakest, which may seem surprising at first sight, since it is the one force that impinges on our everyday experience. A moment's reflection, however, reminds us that we have all seen magnets pick up needles, or an object that has been electrically charged pick up pieces of paper, thereby demonstrating that the electromagnetic force can overcome gravity. The two nuclear forces are not obvious to our everyday consciousness because they are important only over very small distances, comparable in size to atoms and their nuclei. The weak force is involved in the radioactive decay of atoms and the strong force holds together the different particles in atomic nuclei that would otherwise be blown apart by the electrostatic repulsion that the protons in those nuclei would exert on each other.

According to present ideas within the framework of "Big-Bang" cosmology, there was in the very early universe only one force. As the radiant energy in the newly-formed universe cooled sufficiently for matter to appear, the one force "split" into the four we now know. It turns out that, had the relative strengths of these four forces been only very slightly different, life in any form remotely like our own would have been impossible and this has led many to speak of the universe being "fine-tuned" for the existence of life. As Freeman Dyson[8] has expressed it: "The more I examine the universe and study the details of its architecture, the more evidence I find that the universe in some sense must have known that we were coming." As we understand it at present, the ratios of the strengths of these four forces need not inevitably have the values that we have found in this universe. Many people would regard it as intuitively obvious that it is highly improbable for the four forces to have the relative strengths that are needed for the appearance of life as we know it. We must be careful, however, how we use the term "probability". We all know that it is exceedingly improbable that four hands dealt from a randomly shuffled pack of cards will turn out each to have one complete suit. Nevertheless, that particular combination of cards is neither more nor less probable than any other *specified in advance*. We have invested one particular combination of cards with significance, *in advance*, and that accounts for our surprise if it shows up, but it would be just as surprising if someone were to predict accurately the four hands about to be dealt in a game in which that person was playing. Similarly, we cannot help investing with significance those particular ratios of force-strengths that make *our* existence (or that of any intelligent beings) possible; therefore, we are tempted to suggest that it has been specified in advance, although those particular ratios are not inherently more improbable, so far as we know, than any other specified ratios.

It makes sense to talk about the probability of a specified order of cards, because there are so many possibilities – fifty-two multiplied by each of the positive integers less than itself; to be precise, somewhat more than eight followed by sixty-seven zeros – although the odds against each of four hands being a complete suit, while still high, are much lower if we are not concerned about the order of cards in each hand. I am using the term "probability" here in its Laplacean sense: the number of favourable outcomes divided by the total number of possible outcomes. If there is only one universe (a point to which I shall return in the next paragraph) we cannot talk about probability in that way. Such a unique universe is neither probable nor improbable in that sense; it just *is*. There is, however, a sense in which we can talk about probability, as John Leslie[9] has argued in his book *Universes*: suppose that you have the misfortune to be sentenced to face a firing squad of fifty members, each of which is known to be an expert marksman, and you emerge unscathed from the experience. You *may* just have been incredibly lucky, but you and your friends will consider that a much more *probable* explanation is that, for some reason, all the members of the firing squad deliberately aimed wide of you. Whether consciously or otherwise, you and your friends are slipping into Bayesian ideas of probability and attempting to estimate the probability of competing hypotheses from the evidence presented to you – in the fictional case your survival of the firing squad, in the cosmological case the particular kind of universe that has emerged from the Big Bang, but our assessment of that probability is inevitably influenced by the significance with which we invest our own existence.

The situation is different if we believe in the existence of "many universes". The concept of a multiplicity of universes may seem paradoxical; if the universe is defined as "everything there is", there can, by definition, be only one universe. If, however, "universe" is defined as a self-contained domain of space and time in which matter and energy are found, then there could conceivably be many such domains. The use of the word "universe" in this connection is perhaps somewhat unfortunate, and some people prefer to speak of "many worlds", or a multiverse. There are several ways in which we can envisage the formation of such universes or worlds. One of the earliest ways, which is still popular, is based on an interpretation of quantum theory, according to which, whenever two outcomes are possible at the level of fundamental particles, *both* occur and the universe splits into two parallel "universes", each possibility having been realized in one of the resulting universes. If this is correct, there obviously must be an immensely vast number of universes by now and we can again talk in terms of Laplacean probability. (We can only admire the perceptiveness of Hume who, centuries before there was any scientific evidence for multiple "universes", could speak of many worlds being botched and bungled... ere this system was struck out.) In Laplacean terms, life-sustaining universes may indeed be highly improbable, in the sense that their number will be only a very small fraction of the total of all universes that ever have existed or will exist, but it is no longer a cause for surprise that we find ourselves in one of them; we must, after all, be in one of the few universes in which we *can* exist.

Most of us, however, feel instinctively that there is a problem here that needs to be solved and it is very tempting to take anthropic arguments one step further, making them into a version of the argument from design. In this way, one would argue that the universe is the kind of universe in which intelligent life can emerge because whoever, or whatever, created it deliberately intended that intelligent life should emerge. Nevertheless, we must consider all possible explanations and I can think of three besides the possibility of deliberate design:

> 1. The whole universe and everything in it, including you and me, is one tremendous fluke.
>
> 2. There is some law of nature, which we have yet to discover, that determines how the original unified force splits into the four we know, with their relative strengths as we know them.
>
> 3. Some version of the many-worlds hypothesis discussed above is correct and it is almost inevitable that there should be some universes in which intelligent life emerges.

The first of the above possibilities seems to me so unlikely that we can dismiss it. I am tempted to remark that anyone who can believe that will believe anything. I know, however, no cast-iron argument against our universe being a fluke. If some people choose to believe that it is so, that is their privilege; but they are in no position to claim that they are more rational than religious believers.

The discovery of some law of nature that would fix the relative strengths of the four forces is a real possibility, given the attempts under way to find a "theory of everything". It seems to me, however, that such a discovery would only push the problem further back. Why should the laws of nature be such as to permit (or even to demand) the emergence of intelligent life in the universe?

The third possibility, the existence of many worlds, has, as I have remarked above, many supporters, including many, but not all, who have looked carefully into the implications of quantum mechanics. The idea has been around for some time, and an excellent discussion of how it appeared two decades ago can be found in John Leslie's book *Universes* already cited; he updated and simplified this discussion in a more recent book, *Infinite Minds*[10]. Recent developments that bring together our theories both of the very early universe and of the fundamental particles of matter have directed still more attention to the idea of a multiverse and a variety of views on the subject can be found in a recent compendium, *Universe or Multiverse?*, edited by Bernard Carr[11]. Although Carr himself is an enthusiastic proponent of many worlds, as is shown in another paper written together with George Ellis[12] (an opponent), he has assembled a wide variety of views on the subject. The quantum-theory version of the many-worlds hypothesis has those worlds existing concurrently; it is also possible to envisage many worlds existing consecutively. If the expansion of the universe were to slow down and eventually reverse itself, the universe might collapse in a "Big Crunch", only for a new, and perhaps different, Big Bang to set off another universe. That idea had some currency a few years ago when such a cycle of bangs and crunches seemed possible. Our present best knowledge is that the expansion of the universe, so far from slowing down, is accelerating, thus making this particular version of the many-worlds hypothesis seem less probable. If any version of the many-worlds hypothesis is accepted (and there are still others), many believe that the appearance of intelligent life becomes all but inevitable and there is no need to invoke design. We should note, however, that Leslie argues that the conditions for the emergence of intelligent life are so restrictive that it is surprising that there is even one universe in which such life is found – unless, of course, the number of universes is infinite.

Indeed, one motive for adopting the multiverse idea has been to avoid the implications of design inherent in the anthropic principle, but one can argue, with Peacocke[13], that all those universes, even if they are infinite in number, were brought into being in order to ensure that intelligent life would appear sooner or later, somewhere or other. While such a riot of creativity might seem a rather wasteful way of going about things, Robin Collins[14], contributing to Carr's compendium, has argued that just such creativity can be expected of God, and that belief in God is perfectly compatible with the existence of a multiverse. Indeed, Paul Davies[15], writing in the same compendium, argues that the multiverse hypothesis is simply a form of deism. I, personally, am not convinced by the arguments for a multiverse, but I freely admit that my own reaction is not scientific, nor even logical, but aesthetic. The concept of a multiverse does not ring true for me, although, as I mentioned in Chapter 1, Buddhists or Hindus could find it quite consonant with their religious beliefs. Cosmology is an area in which even great scientists have allowed their aesthetic preferences to influence their scientific opinions, so I feel free to follow in their footsteps, especially since expert cosmologists have not yet reached a consensus on the multiverse.

The brief discussion by Carr and Ellis, already cited, provides a good summary for those who would like to pursue the matter further, although many of the arguments that those two discuss raise more technical issues than I wish to go into here. Ellis places considerable weight on the argument that multiverse theory is not testable because it can be used to explain anything at all and is therefore not a part of science as it is commonly understood. Carr argues that our understanding of the nature of science has changed and will change again and that the multiverse theory unites work at the very largest scales and at the very smallest (the universe itself and fundamental particles). The two look at the same data, about which they largely agree, but interpret them differently, thus, incidentally, illustrating Polanyi's point that scientific knowledge is not completely objective[16].

Paul Davies[17] has recently put forward yet another possible explanation for the so-called "cosmic fine-tuning" in his book *The Goldilocks Enigma*. That explanation is based on suggestions by Stephen Hawking and J.A. Wheeler, who argued that, at the quantum level, the nature of the past is determined by the observations that we choose to make in the present. An example is the famous two-slit experiment by which Thomas Young, in the early nineteenth century, demonstrated conclusively, as it then seemed, that light consisted of waves rather than particles. Light from a single source is passed through two parallel slits and focussed on a screen. An interference pattern of alternate light and dark bands is built up on the screen. We now know that the pattern can also be explained in terms of the particle picture of light, so long as we are dealing with large numbers of particles (photons). Suppose, however, that the experiment is designed so that only one photon at a time passes through the apparatus. In course of time, an interference pattern will ultimately appear, but which photon has passed through which slit? You cannot determine this without destroying the interference pattern. Even if you delay the attempt to determine the path of any given photon that helped to produce an already-existing interference pattern, you still destroy the pattern. Davies argues that your observation determines the nature of the past. In this experiment, one is talking of only a few nanoseconds back in time, but Davies argues that, in principle, one could extend this sort of argument back to the Big Bang itself, and that the existence of life in the universe now has in some sense determined which past history of many possible ones our universe actually experienced. This is, of course, completely counter-intuitive – as much of quantum physics is – and is also highly speculative, but some people, at least, take it seriously.

It must, however, be clear to the reader by now that the explanation that I most favour for the "fine-tuning" of the universe is that the universe was created with the deliberate intention that intelligent self-aware life should emerge. I do not claim, however, that the observational data lead inevitably to that position. I am not so much advocating that the universe is designed and therefore God exists as saying that because I already believe that God exists, I can easily accept the notion that the universe has been designed. A similar view has recently been expressed by Owen Gingerich[18]. What we are all doing, of course, whichever explanation of the fine-tuning we prefer, is assigning Bayesian prior probabilities to them all. The probabilities we assign are influenced by the beliefs (or unbeliefs) that we already hold, and, as I have already pointed out, the way we look at new evidence is also subtly influenced by those same beliefs. That is all right so long as we acknowledge the fact and try to be open to the possibility that new evidence may one day force us to revise our beliefs. In the meantime, we should refrain from labelling as "irrational" those who assign probabilities that differ from our own. I am not sure that, on average, scientific materialists are any better than religious believers at meeting those conditions.

The real strength of anthropic arguments, it seems to me, is that they remind us that we are part of the universe that we study. Science has been built on objectivity. We scientists study objects, whether they be stars, rocks, chemical elements, or plants and animals, that we think of as quite separate from ourselves. We try neither to affect them by our studies, nor to be affected by them. Although, as we have seen, Polanyi has argued that even the most exact sciences cannot be completely objective, this kind of objectivity works very well over a wide range of object sizes, say from a grain of sand to a cluster of galaxies. When we look at the very small (fundamental particles), however, we find that our very attempts to observe them affect them, and when we look at the universe as a whole we cannot separate ourselves from it. (Scientific objectivity is also harder to maintain when the objects of study are living things, and even more so when they are fellow human beings, but that is a rather different matter.) The kind of argument put forward by Bertrand Russell and quoted in the Prologue is then seen to be based on a misconception. Russell saw humanity as standing alone in defiance of a universe that is largely hostile to life, but we are part of the universe, the very special properties of which favour, or possibly even ensure, the appearance of life, intelligence and, ultimately, moral awareness. We could not live either in stars or in the vast spaces between them that so terrified Pascal; both, indeed, are hostile to life, but the stars are necessary for building up the heavy elements found in our bodies, and the interstellar spaces are necessary for the stars to develop as they do. People who remark that they can find no evidence of purpose in the universe forget that they themselves are part of it. In us, *if nowhere else*, the universe has become self-aware, purposeful and morally conscious. This may sound like anthropocentric hubris, but I shall discuss in the next chapter that important qualifying phrase in italics.

The history of the design argument in the biological sciences is different, though not without parallels. Biologists before Darwin were neither blind nor stupid, and were fully aware that most animals are well adapted to their respective environments, to what we would call their ecological niches. Those early biologists saw the facts of adaptation as strong evidence for the wisdom of the Creator. Although, as we saw in the last chapter, natural theology has a long history within Christianity, during the late eighteenth and early nineteenth centuries it became particularly associated with this form of the argument from design. Before judging those biologists and theologians, we should recall that virtually everyone in the Western world at that time believed in the fixity of species and in what John Ray called "the novity of the Earth". The two go together, of course, because, as we have seen, Darwin required the long periods of time that the geologists argued for, if evolution by natural selection was to be a tenable hypothesis. Until Hutton and Lyell provided those geological arguments, theories of evolution were bound to remain the sort of speculative and poetic ideas that Darwin's own grandfather put forward, and it was not unreasonable to believe that species were created, in pretty much their present form, at a fairly recent beginning. At that time, natural theology was as much a theory of the origin of adaptations as the theory of natural selection that replaced it is now. The argument that both "theories" should be taught together could have carried much more weight in the mid-nineteenth century than it should today.

That being said, it must be admitted that some eager natural theologians pushed their argument to extremes. According to Bertrand Russell[19], some of them even went so far as to claim that it was providential that rabbits had white tails that provided a clear target for sportsmen! (Russell does not give his source for this statement, and he sometimes stretched the truth in order to make a witty point.) Even in its more restrained form, however, the theological argument was based on a static origin for adaptations and was, ultimately, another example of the God of the gaps. Darwin undercut that kind of natural theology by showing that adaptation could be explained without appeal to Divine intervention, by the *dynamic* mechanism of natural selection. The eclipse of natural theology, to which I alluded in the previous chapter, may well have been partly due to the fact that theologians themselves realized that in its more extreme forms it was untenable. The mechanism of natural selection is the sticking point for many who did, and still do, oppose the idea of Darwinian evolution. Indeed, Darwinism itself fell out of favour, after its initial surge of popularity, until about the 1930s precisely because of doubts, which were as much scientific as religious, as to whether natural selection could explain the origin of species and their adaptations (see Sir Alister Hardy[20], *The Living Stream*, for a more detailed history of this episode). Here, again, we may see one reason why so many modern evolutionists insist so strongly that natural selection is all that is needed. They are not only fighting against fundamentalist religious believers, but also still fighting the scientific battles of earlier generations.

Darwinism emerged from its eclipse after the rediscovery of Mendel's laws of heredity and the consequent development of genetics, which showed how natural selection might change species. By the late 1930s, the two strands of Darwinism and Mendelism had been woven into what Julian Huxley termed the "modern synthesis". Shortly after, as we saw in Chapter 1, DNA was identified as the likely carrier of genetic information and when, in 1953, Watson and Crick discovered its molecular structure, the method by which the molecule replicated itself immediately became obvious, and the way in which mutations might arise, on which natural selection could operate, was revealed. Although Ronald Numbers[21] sees the rise of modern "creationism", initially in the United States, as a reaction to the revisions and strengthening of American school science curricula after the successful launch of the sputniks by the Soviet Union in 1957, reaction to the apparently invincible progress of molecular biology, which seems to so many of its practitioners to support an entirely materialistic interpretation of life and has confronted us with so many ethical problems, must surely be a factor as well..

If the basic ideas of Darwinism were false, it is unlikely that so many applications of modern genetics would be as successful as they are, so the above-mentioned ethical controversies arising from these applications help to show that Darwinian ideas are as nearly certain as any scientific ideas can be. Even if the modern synthesis is destined to be superseded some day, the new theory will be a yet more far-reaching one, containing Darwinian ideas within it, in much the same way as Newtonian mechanics is contained within Einstein's theory of relativity as a special case. Despite some vigorous argumentation by Alvin Plantinga[22], we can no more go back to notions of the special creation of each species and a young Earth than we can to a geocentric universe. Yet, just as physical design arguments have been reborn in the guise of the anthropic principle, biological design arguments have been reborn in the shape of "Intelligent Design".

Although Phillip Johnson introduced the concept of Intelligent Design to account for the creation of new species and creationists have seized on such arguments as support for their own position, it is perhaps unfair to lump all proponents of Intelligent Design together with creationists. Two prominent advocates of Intelligent Design, Michael Behe[23] and William Dembski[24, 25], accept the astronomical and geological evidence for the age of the Earth and the biological evidence for the origin of species by evolution. Their arguments for design concern rather the origin of various structures within whole groups of organisms – all mammals, or even all vertebrates, for example – which they believe could not have arisen by a gradual process such as natural selection. As Eugenie Scott[26] has pointed out, these arguments have a much broader appeal than has outright creationism. Indeed, accepting evolution and an old Earth while insisting that design played a role in the former probably seems a sensible compromise to many non-biologists, although the overwhelming majority of evolutionary biologists, including many who openly identify themselves as

Christians, consider the hypothesis of Intelligent Design to be both unnecessary and mistaken.

Intelligent Design would probably be much less controversial than it is if there were no campaign to have it taught in high schools as a fully developed alternative theory to Darwinian evolution. Such a course of action would undoubtedly be wrong, given the overwhelming consensus referred to above. In the absence of the campaign, Darwinians would, no doubt, still object that Intelligent Design was an unnecessary theory, but they would probably largely ignore it, regarding it as fringe science, unworthy of serious consideration. As it is, Intelligent Design is hardly entitled to the status of a coherent alternative theory; it is an assertion that some "irreducibly complex" structures that we find in living creatures could not have arisen by the gradual process that natural selection inevitably is, because, so it is claimed, those structures would be of use to the organisms in which they are found only if they were complete.

Even a superficial knowledge of physiology (I claim no more myself) leads anyone of us to marvel at the construction of our own bodies and, by extension, that of all living organisms, even those consisting of a single cell. I think that many who have not through their own work become familiar with the powers of natural selection have genuine difficulties in accepting that that mechanism can explain some of the more complex organs and processes within our bodies, even though they completely accept the idea of evolution. Evolutionary biologists and geneticists perhaps do not always fully appreciate how those difficulties appear to others, and some outlines of explanations might do more to convince many people than rather polemical attacks on Intelligent Design. In this context, it is perhaps noteworthy that neither Michael Behe nor William Dembski are evolutionary biologists but, respectively, a biochemist and a mathematician; each, as it happens, is also a Christian with a religious motive for arguing for design. Science has sometimes progressed when individuals have ventured across traditional disciplinary boundaries, and it is perhaps fair to suggest that attempts to refute Behe and Dembski have at least stimulated fresh work.

All parties agree on the wonderful, even awe-inspiring, complexity of living organisms; that many features of living organisms *appear* to be designed; and that Darwin himself was well aware of arguments like those put forward by Behe and Dembski, as is shown in a section of *The Origin of Species*[27] entitled "Organs of Extreme Perfection". He went so far as to acknowledge that if opponents of his theory could show definitively, of even a single organ, that it could not be produced by natural selection acting on random variations, then his whole theory would fail. Darwin was confident, however, that such an example could not be found, and that is where disagreement emerges. The mainstream of evolutionary scientists is certain that subsequent research has vindicated Darwin, while Behe and Dembski insist that there are at least some structures within living organisms that not only appear to have been designed, but actually have been.

Few organs, if any, are of "extreme perfection". One example, often cited long before the term "Intelligent Design" was coined, is the eye of human beings, and other vertebrates (insect eyes are built on a different pattern); but consider how many of us need additional lenses to correct the defects of our natural ones! Indeed, by late middle age, almost all of us need some help. Once again Darwin saw that defects in the organ would be expected of evolution by natural selection, whereas most of us instinctively feel that an Intelligent Designer ought to have done better. Both Behe and Dembski concede that much of what they perceive as design is imperfect. Indeed, Behe actually compares some of the features he considers to be designed to Rube Goldberg machines and he also argues (pp. 222-225 of his book) that the Designer may have had other motives than perfection in design, that we do not and maybe cannot understand – an argument that seems to me seriously to compromise the claim that Intelligent Design is a scientific theory, since it tends to stifle further investigation. Dembski, on the other hand, stresses that the Designer need not necessarily be the omnipotent and loving God of Christianity. In the last chapter of *No Free Lunch*, he often refers simply to an "unembodied designer" – which perhaps could be something like Plato's demiurge, making the best use of the material available. While such a concept cannot be logically ruled out, it seems to me that it would raise at least as many problems for biblical literalists as does evolution by natural selection.

The eyes of vertebrates are sometimes called "camera eyes" because their mode of operation is very similar to that of a camera. Doubts that such eyes could be the result of natural selection arose because their functioning depends on several different parts working together, most of which, individually, do not confer any obvious advantage on the creatures that possess them. The eye needs a lens, an iris, a retina, an optic nerve to convey signals to the brain, and has two compartments containing fluids of different refractive indices. It is not immediately obvious that each of these by itself would give a creature much advantage. Thus, it was argued that natural selection had nothing to work on and the whole eye must have been created at one go. In fact, geneticists are virtually unanimous that natural selection *can* explain the evolution of the camera eye and that it appears to have evolved independently several times, in different lineages of species. Behe, however, although he mentions the anatomical structure of the eye, is more concerned with molecular structure of the various light-sensitive molecules that send the signals that our brains transform into vision (this comes out more clearly in a paper on his website than in his book). Unsurprisingly for a biochemist, Behe places quite a lot of emphasis on molecular shapes and the close fit between certain molecules and other receptor molecules. Kenneth Miller[28], in his book *Finding Darwin's God*, points out that molecules and their receptors can evolve together, thus developing the close fit. Moreover the properties of molecules, including their shapes, are at least partly determined by the atoms that make them up, and the forces acting between those atoms. Much of Behe's argument, therefore, reduces to a special case of the anthropic argument, and is not directly relevant to the question of the adequacy or otherwise of natural selection.

Dembski's approach is different; he believes that he can specify mathematically when the appearance of design is, in fact, more than just appearance. While he makes use of Behe's examples, his principal argument is a mathematical one and might still be valid even if all of Behe's examples are fallacious. His formula has been disputed by, for example, Wesley Elsberry[29]. I have not myself been fully convinced that Dembski has made his case, but his arguments are difficult for anyone to assess who does not have his obvious familiarity with modern statistical and complexity theory. As already mentioned Dembski does not insist that the biological design that he believes he has demonstrated is necessarily the work of a supernatural designer. Rather, in a paper on his website he argues that mind is as much a part of the natural universe as is matter and he presents "Intelligent-Design theory" as simply an attempt to include a role for mind in the natural universe. That is a point of view with which I have some sympathy, but it raises the whole question of the relation of the mind to the brain, which I will try to discuss in Chapter 6. Those who believe in the identity of the mind and the brain will, of course, immediately dismiss the notion that mind could in any way affect evolution, but not all Darwinians would take that position and I do not see that Darwinism must necessarily exclude a role for mind.

Of all the examples of "irreducible complexity" that Behe adduces, the one I found most convincing was the mammalian blood-clotting cascade, which he believes could not have arisen by the gradual process of natural selection. However, Miller has taken up Behe's challenge in *Finding Darwin's God*, and explained how the blood-clotting mechanism of vertebrates *could* have arisen step by step, giving the organism gradually increasing protection against unlimited bleeding – an explanation that he has elaborated on his website[30]. Miller points out that some invertebrates (specifically lobsters) have a kind of blood-clotting mechanism, thus undercutting Behe's claim that the mammalian system is irreducibly complex.

Blood-clotting is but one example of many bodily processes (including breathing and digestion) that proceed without any conscious effort on our part and of which we remain unaware except when something goes wrong with them. The mechanisms that maintain our bodies at a more-or-less constant temperature, or keep our blood sugar within safe limits, are things that most of us never give a thought to – unless we belong to that unlucky minority in whose bodies one or other of those homeostatic systems do not work. I, personally, find these forms of homeostasis a greater challenge to natural selection than the supposed perfection of the eye; but I do not venture to argue that they cannot be explained by that means. To do so would be using what Richard Dawkins[31] has called the "Argument from Personal Incredulity" (his capitals), and I believe that he is quite right to deny its validity. After all, many Christians would, in their turn, deny that Dawkins' inability to believe in the Resurrection is a valid argument against it. The very fact that these homeostatic systems sometimes break down makes them further examples of imperfect design. One would expect a truly Intelligent Designer to have designed a fail-safe mechanism, or at least a back-up process in case of need. The vulnerability of homeostatic systems in individuals under stress for some reason or other is, on the other hand, exactly what one might expect from the trial-and-error process that is natural selection.

In his book *The Tinkerer's Accomplice* J. Scott Turner[32] also discusses homeostasis, although he is not so much concerned with particular homeostatic mechanisms in mammalian bodies as with the interaction between whole organisms and their environments. As a physiologist, he sees natural selection working at the level of the whole organism – over against Dawkins' insistence on the "selfish gene". Individual organisms can, in Turner's view, react on the environment, tending to make it more suitable for themselves. This aspect of homeostasis is the accomplice of natural selection – the "tinkerer" – and the results of their interactions appear to us as design. Turner's approach is thus quite different from that of the proponents of Intelligent Design. Indeed, he points out that they have to some extent missed the boat by concentrating their attention on things like the eye, rather than on the processes in the brain that convert the image on the retina into (for the most part) a remarkably accurate three-dimensional plan of the world we move about in. He describes those processes at some length (as does Crick in *The Astonishing Hypothesis*) and indeed they provide another example of our bodily workings that are to be marvelled at. Yet, strangely enough, they do not strike us as designed – most of us imagine that any human designer sitting down to solve that particular problem for a robot would come up with something much simpler.

Because Turner sees organisms modifying the environment for their own benefit, he allows the environment to have at least an indirect effect on their offspring. The central dogma of molecular biology, that information passes only from DNA, through RNA, to protein, which rules out the Lamarckian inheritance of acquired characters, is not always the one-way street it was initially assumed to be. Turner sees organisms contributing to their own evolution by the effects that they have on the environment. Perhaps this is the way in which mind can play a role in evolution, such as Dembski wishes for; although, of course, it is mind operating at a level far below the self-consciousness it reaches in human beings. Yet Turner sees this contribution of organisms to their own evolution as in some sense intentional and even suggests that that intentionality goes back to the origin of life itself. Although he never mentions James Lovelock[33] and Gaia, there is a parallelism between the thinking of the two, in that Lovelock sees life, once it is established, working to preserve the environment in order to keep it hospitable. Strangely, there is even another parallel between Turner and the suggestions by Davies discussed earlier in this chapter, in that both argue that what exists now could, in some way, have helped to create the conditions needed for its existence. Here are two mainstream scientists, working in quite different areas of science, and, therefore, looking at different phenomena and using quite different arguments, coming up with surprisingly similar conclusions. Although neither uses these words, we could deduce from what they say that life is the manifestation of something that has shaped the world we live in, perhaps even the universe itself. I hesitate to replace the word "something" with words like "force" or "power" since for scientists, at least for physical scientists, those words stand for precisely defined technical terms. Those for whom the words have a more general connotation, however, are free to adopt them, if they find them helpful.

A final comment about design in the context of biology is that we have to be careful from whose point of view we are speaking. A freshly spun spider's web is, to most of us, an object of beauty and an outstanding example of design in nature. When light from the low Sun glances on such a web, it shimmers in various colours in much the same way (and for the same reason) as does the playing side of a compact disc seen edge-on. When the Sun shines directly through such a web, it appears, both literally and metaphorically, in a different light. The body of the spider is revealed, as are the bodies of the flies and other insects that its web has caught. We are reminded that what appeared as a thing of beauty at another time is, from the flies' point of view, a terribly efficient weapon of mass destruction. If flies could reason, they might well see the web as a product of design – but a diabolical design rather than a design of a beneficent deity interested in creating beauty. In the flies' theology, cobwebs would be a major component of their problem of evil. There is, of course, a third point of view: that of the spider. I suppose the spider never sees its own web as a whole and, even if it were capable of any kind of aesthetic appreciation, would never regard the web as an object of beauty. At some level it must be "conscious" of the web as an extension of its own body – from which, indeed, the material of the web did come – and a necessary means of obtaining food. Which of these three points of view is the correct one?

From the theological point of view, however, the chief objection to design arguments remains that the God whose existence they purport to demonstrate is a "God of the gaps". Newton thought the design of the solar system was proof of Divine Providence; a century later, Laplace had "no need of that hypothesis". Dembski believes that he has devised a mathematical way of overcoming this objection, but my reading of the history of science makes me cautious about accepting his claim. Anthropic arguments look strongly suggestive but if, despite my own reservations, some form of the many-worlds hypothesis is shown to be true, even those arguments will fail to convince. Eighteenth-century theologians thought adaptations in living creatures to be strong evidence for the existence of God, evidence that Darwin undercut completely.

Intelligent Design, if it is indeed intended as an argument for the existence of God, seems to me particularly vulnerable. Let us grant that there are some features of our bodies that *at present* are "irreducibly complex"; those will be the very features that geneticists will select for concentrated study, and will, in all likelihood, succeed in showing that they could be produced by natural selection. For example, Miller's outline of how the blood-clotting mechanism could have evolved gradually seems to have been stimulated precisely by Behe's challenge. Then a new gap has to be found for God. Like the other four ways of Aquinas, design arguments can help to show that it is not totally irrational to believe in God, but they fall short of being conclusive proofs.

References:

[1] Hume, D., *Dialogues Concerning Natural Religion*, 1779, many reprints e.g. *Hume on Religion*, ed. Richard Wollheim (Fontana Library, Collins, London, 1963), pp. 99-204.

[2] Newton, I., *Principia Mathematica*, 1687. (The text reference is to the *General Scholium* found in the 2nd edition – 1713. A revision of the English translation by Andrew Motte was made by Florian Cajori and published posthumously by the University of California Press, Berkeley, California in 1946. In that edition, the *General Scholium* is on pp. 543-547.)

[3] Laplace, P.S., 1796, *Exposition du système du monde*, Paris.

[4] Kant, I., *Allgemeine Naturgeschicte mit Theorie des Himmels*, 1755, English translation by W. Hastie, 1900, reprinted as *Universal Natural History and Theory of the Heavens by Immanuel Kant*, with an introduction by M. K. Munitz (Ann Arbor Paperbacks, Michigan University Press, 1969), Second Part, First Chap., pp. 71-82.

[5] Lubbock, C. A., *The Herschel Chronicle* (Cambridge University Press, 1933), pp. 309-10.

[6] Foundling Hospital Hymn Collection, 1796.

7. Carter, B., in *Confrontation of Cosmological Theories with Observational Data*, ed. M.S. Longair, IAU Symposium, No. 63 (D. Reidel, Dordrecht, Holland, 1974), pp. 291-298.
8. Dyson, F.J., *Disturbing the Universe* (Harper Row Publishers, New York, 1979), p. 250.
9. Leslie, J., *Universes* (Routledge, London and New York, 1989).
10. Leslie, J., *Infinite Minds* (Oxford University Press, 2001), Chap. VI.
11. Carr, B. (ed.), *Universe or Multiverse?* (Cambridge University Press, 2007).
12. Carr, B., and Ellis, G., *Astronomy and Geophysics*, **49**, 2.29-2.37, 2008.
13. Peacocke, A.R., *Theology for a Scientific Age*, enlarged edition (SCM Press, London, 1993), pp. 106-109.
14. Collins, R., in B. Carr (ed.), see reference 11, pp. 459-480.
15. Davies, P.C.W. ibid., pp. 487-505.
16. Polanyi, M., *Personal Knowledge* (University of Chicago Press, 1959; corrected edn. 1962).
17. Davies, P.C.W. *The Goldilocks Enigma: Why the Universe is Just Right for Life* (also published under the title *Cosmic Jackpot*) (Houghton Mifflin Company, Boston and New York, 2006), pp. 242-249. See also his article in *New Scientist*, 30 June 2007, pp. 30-34.
18. Gingerich, O., *God's Universe* (Belknap Press, Harvard University Press, 2006).
19. Russell, B., *Outline of Intellectual Rubbish*, 1943, reprinted in *The Basic Writings of Bertrand Russell*. See reference in Prologue.
20. Hardy, A.C., *The Living Stream* (Collins, London, 1967), Chap. III.
21. Numbers, R.L., in *Scientists Confront Creationism: Intelligent Design and Beyond*, eds. A.J. Petto and L.R. Godfrey (W.W. Norton & Company, New York, 2007), pp. 31-58.
22. Plantinga, A., *When Faith and Reason Clash: Evolution and the Bible*, in *Christian Scholars' Review*, **XXI**, No. 1, 8-33, 1991.

23. Behe, M., *Darwin's Black Box: The Biochemical Challenge to Evolution* (The Free Press (Simon and Schuster), New York, 1996), esp. p. 223. See also *Evidence for Intelligent Design from Biochemistry*, text of a lecture delivered in August 1996, found on the website <http://www.arn.org/docs/behe/mb_idfrombiochemistry.htm>.
24. Dembski, W., *The Design Inference: Eliminating Chance through Small Probabilities* (Cambridge University Press, 1998).
25. Dembski, W., 2002, *No Free Lunch: Why Specified Complexity cannot be Purchased without Intelligence* (Rowman and Littlefield, Lanham, Maryland, 2002). See also *In Defense of Intelligent Design*, in *Oxford Handbook of Religion and Science*, eds P. Clayton and Z. Simpson (Oxford University Press, 2006), pp. 715-731 (also on Dembski's web page).
26. Scott, E.G., see reference 21, pp. 59-109.
27. Darwin, C., *On the Origin of Species...* (John Murray, London, 1859) Many reprints. Chap. 6, pp. 217-224 in the Penguin edition of 1968.
28. Miller, K., *Finding Darwin's God: A Scientist's Search for Common Ground between God and Evolution* (Cliff Street Books (Harper Collins Publishers), New York, 1999), pp. 143-144.
29. Elsberry, W.R., see reference 21, pp. 250-271.
30. Miller, K., <http://www.millerandlevine.com/km/evol/DI/clot/Clotting.html> accessed March 2009.
31. Dawkins, R., *The Blind Watchmaker* (W.W. Norton & Company, New York and London, 1987), p. 38.
32. Turner, J.S., *The Tinkerer's Accomplice* (Harvard University Press, 2007), p. 226.
33. Lovelock, J., *Gaia: A New Look at Life on Earth* (Oxford University Press, 1979).

CHAPTER 5
Life and Consciousness in the Universe

> La marquise sentit une vraie impatience de savoir ce que les étoiles fixes deviendraient. Seront-elles habitées comme les planets, me dit-elle? Ne le seront elles pas? Enfin qu'en ferons nous?
> Bernard le Bovier Fontenelle, *Entretiens sur la pluralité des mondes*

In the last chapter I risked saying that in us, if nowhere else, the universe has become self-aware, purposive and morally conscious. I talk of "risk" because I am well aware that the statement may be misunderstood as implying that I believe the universe to have been created solely for the benefit of the human race. In fact, I agree with Dean Inge's[1] remark "There is, I think, something derogatory to the Deity in supposing that he made this vast universe for so paltry an end as the production of ourselves and our friends." Inge wrote that in the year that I was born and I emphasized in the first chapter how greatly our conception of the vastness of the universe has increased in my own lifetime. He went on to question whether the universe had one "infinite" purpose or, as he thought more likely, a series of finite purposes. I am not quite sure why he chose the terms "infinite" and "finite" but I take it that he meant to imply that the human race might be one of many purposes that the God in Whom he believed had in creating the universe. I would go even further and suggest that we might be merely a by-product of some much greater purpose. Whether or not there is a sense in which we are made "in the image and likeness of God", we are not the unique crown of all creation that we used to like to think we were.

Such thoughts as these inevitably lead us to consider the possibility of life elsewhere in the universe. Although there were speculations about that possibility in antiquity, Copernican ideas undoubtedly stirred up a greater interest in it, as they became more widely known. In very ancient times, the planets were identified with the gods whose names they bore and divine and semi-divine beings were thought of as populating the heavens. Then Christians placed Heaven in the heavens and populated them with angels and the souls of the blessed, but that was not quite the same thing as imagining other mortal creatures of flesh and blood inhabiting other worlds. Copernicanism brought home the fact that there *were* other worlds (in the sense, here, of other inhabitable planets, not in the sense of other universes used in the last chapter).

Perhaps a short digression on the subject of angels is desirable, although I shall discuss the broader question of the role of imagery in science and religion in Chapter 8. In our modern age, many people regard the whole idea of angels as nonsense, although some of the same people have no difficulty in believing that little green men from other planets have visited us and even abducted some of our number. On the other hand, there are many literally-minded believers who maintain that angels exist and are exactly like their traditional representations. This paragraph will please neither group. Of course, the traditional imagery *is* nonsensical. Humanoid figures in long robes, equipped with a pair of wings sprouting from somewhere in the neighbourhood of their shoulder blades (even great artists usually avoid showing just how the wings are attached!) would be totally unstable in flight, even if they could take off. That in itself, however, does not make it irrational to be open to the possibility that there are higher orders of being than we commonly encounter unless, of course, one is committed to the belief that the world of the senses is all there is and what I have called the transcendent does not exist. Every time I dig in my garden, I cause a catastrophe for earthworms; something comparable, on their scale of being, with the effects on our race of the tsunami that occurred in the Indian Ocean a few years ago. I literally turn the world of the worms upside down. I kill or injure many directly with my spade, and others indirectly by making them easy prey for our so-called American robins – which share with the true robins not only a red breast, but a readiness to approach gardeners while they are digging, for the sake of a juicy worm. If the surviving worms could reflect on the experience, however, they could be, at most, only dimly aware of the being who had caused all this mayhem. I do not claim to have had any experiences of angels, but I do not see it as a total impossibility that non-physical entities, as far superior to us as we are to the worms, should exist.

Let us return, however, to the possibility of life existing elsewhere in the physical universe, in forms sufficiently like our own that we would be able to recognize the creatures endowed with it as being alive, and with whom we might, even, be able to communicate; and let us confine our attention to this kind of life. As we have seen, the Copernican revolution provided a stimulus to this sort of speculation, even though the Church frowned upon the idea. Johannes Kepler[2] (1571-1630) published his dream of a voyage to the Moon, while in 1686 Bernard le Bovier de Fontenelle[3] (1657-1757) published *Entretiens sur la pluralité des mondes.* These early speculations were not informed by any of our modern scientific understanding of the conditions necessary for life and Fontenelle's book seems quite fanciful to the modern reader. As the quotation at the head of this chapter shows, Fontenelle made no distinction between the planets and the fixed stars when considering the possibility of life elsewhere in the universe. He did not think in terms of planets circling other stars. Even a century later, no less a person than Sir William Herschel (1738-1822) thought it possible that the bright "surface" of the Sun that we see was composed of luminous clouds, under which there might be a solid surface on which solar beings lived.

Early writers on the subject, especially those who lived in countries where the writ of the Roman Church still ran, had to be circumspect in what they wrote. If there were people even on the nearby Moon (something still thought possible in the seventeenth and eighteenth centuries) how could they be told the good news of Christ's redeeming sacrifice, given that no means of communication with them could be envisaged? This kind of objection lasted until our own day. Early in my own career, a senior colleague who was also a devout Roman Catholic confided that he had not felt fully free to give popular talks on the subject of life on other worlds until the time of the Second Vatican Council. Even E.A. Milne[4], an outstanding theoretical astrophysicist and cosmologist, suggested not long before he died in 1950 that the then unidentified radio sources in the sky were signals from other civilizations and therefore it could be possible, in principle, for us to communicate the Christian Gospel to them. Historically, considerations like these inhibited discussion for a long time, but they are no longer considered important by most theologians, if by any. It is of some interest that a recent survey reported by Ted Peters[5] shows that non-believers are more inclined to think that the discovery of life elsewhere in the universe will have a disturbing effect on religious belief than are those who identify themselves as believers.

In our time, we know much more both about the conditions needed for life and about the conditions to be found on other bodies in our solar system and we can evaluate the possibilities in a much more realistic way. Even by the late nineteenth century, ideas such as men in the Moon or Herschel's solar creatures could be ruled out. At that time, nothing was known about planets around other stars, but astronomers were beginning to understand something of the nature of the other planets in our system. To use modern terms, the terrestrial planets (Mercury, Venus, Earth and Mars) seemed the most promising locales, with attention being particularly focussed on Venus and Mars, since Mercury, like our Moon, has no atmosphere. In the early twentieth century, many thought the existence of Martian beings to be quite possible; Percival Lowell, convinced by his own observations, engaged in sufficient publicity to ensure that the idea was widely known, while H.G. Wells'[6] *War of the Worlds*, although admittedly science fiction, seemed all too plausible. Even when I began my career, Venus and Mars were still thought to be the most likely places in the solar system where life of some sort might be found. Soon after, however, space probes showed us how hellishly hot the surface of Venus is, and we no longer look for life there. The probes sent to Mars, on the other hand, have left us in doubt; no sure sign of life has yet been found, but the planet does have an atmosphere and the existence of frozen water on the planet has been confirmed while this book was being written (and it seems likely that at some time in the past there was liquid water there). Conditions on Mars now, however, are so extreme that we can scarcely hope to find anything above the level of bacteria. The same age of space exploration has taught us to look again at the larger satellites of Jupiter and Saturn; many believe that we shall find signs of life there, but, again, only simple forms. We cannot expect to find intelligent life anywhere in the solar system but here on Earth.

Of course, well before the space age it was recognized that other stars might be accompanied by systems of planets. Indeed, if the modernized version of Laplace's theory of the origin of the solar system is correct, almost all stars like the Sun – of which there are very many, even in our own Galaxy – will have an attendant system of planets. Until recently, however, we could argue only in such general terms. For the most part, planets around other stars are not seen directly, but detected by the effect they have on the motion of their parent stars. This effect is tiny and only recently have the techniques required to detect it reached the necessary precision. In the past decade or somewhat more, however, over 400 planets have been discovered revolving around other stars; some of those stars have two or more planets revolving around them, but the total number of such planetary systems exceeds 200 (at the time of writing, the numbers are continually increasing). Most of these systems do not help us very much in our attempts to assess the likely frequency of life elsewhere in our Galaxy because they have turned out to be very different from our own solar system, with planets several times the size of Jupiter revolving around their parent stars in periods like that of the Earth, or even shorter. There is unlikely to be room in such a system for an Earth-like planet at a suitable distance from its primary star for advanced forms of life to develop on it.

The discovery of these different kinds of planetary system was a surprise to those who work on theories of the origin of our own, but is not, at least in retrospect, surprising in itself. Given that such systems exist, they are easier to detect than are ones more like our own, both because the effect of the very large planets on the motion of the parent star is larger and because the shorter periods enable us to determine more quickly the nature of the perturbation in that star's observed motion. If we were observing our own Sun from a planet revolving around some other star, we would have to wait at least eleven years to confirm that the effect of Jupiter on the Sun's motion was indeed a periodic one, and the size of that perturbation would be several times less than those of the first discovered planets around other stars. That we have not observed systems like our own does not mean that they do not exist, but we still do not know how common they are.

We seek other Earth-like planets, but how like the Earth do they have to be? If life is to take anything like the course it has taken here on Earth, the planet needs air, water and land. Only a sufficiently massive planet will retain an atmosphere – in our own system, Mercury has none and Mars only a very thin one – and only a planet neither too close nor too far away from its parent star will have water in liquid form. On the other hand, a planet much more massive than the Earth, even if it is not a "gas giant" like Jupiter or Saturn, will have a correspondingly greater force of gravity and that will set limits on the size of skeletons that can develop in land animals for any given bone material. Although some very large dinosaurs showed that large animals can live on the land, it is no accident that the largest mammals we know, the great whales, are those who have returned to the sea, the buoyancy of which supports their weight. These considerations, therefore, limit our search to planets not so very different from the Earth, at distances from their parent stars not greatly different from that of the Earth from the Sun. At the time of writing, no such planet has been detected, or, at least, no account of one has been published. The precision of techniques of detection, however, has increased to the point that it is possible, in principle, to discover such a planet, and we know of a planet around the star Gliese 581 that is only a few times more massive than the Earth, and is at the right distance from its parent star to have liquid water. At present, however, we know nothing about the kind of planet this may be – whether it is a large Earth-like planet, or a smaller version of Neptune in our system. Techniques for detecting planets are being developed rapidly, and the situation may change (literally) overnight.

There are other considerations that would be harder for us to determine from a great distance. The Earth's magnetic field plays a part in making life possible here by protecting us from streams of charged particles emanating from the Sun. At times of intense solar activity we see some of these particles as aurorae, but most of the time they are deflected away from us. A constant stream of these particles could be damaging to life. The presence of the Moon may also be important for the development of life. Without the Moon, tides in our oceans would be about one-third of what they actually are. The tides have certainly played a role in the way that life has developed and may even have played a role in the way that life began. Darwin envisaged life beginning in a "warm little pond"; the pond might have been a tidal pool. The suggestion has also been made that the Moon plays an important role in keeping the tilt of the Earth's axis to its orbital plane within narrow limits, which is vital for the stability of the seasons. The importance of this role has recently been questioned by David Waltham[7]; the Earth's axis could also be kept stable if the Earth rotated more rapidly. Perhaps, however, too rapid a diurnal motion would make the development of life more unlikely, so the Moon may be important because it keeps the Earth's rotation relatively slow. In our own solar system the Moon is unique, not in being the largest satellite, but in being (if we exclude the recently demoted Pluto and its satellite Charon) the satellite most nearly comparable in size to the planet around which it revolves. The Earth-Moon system has sometimes been described as a double planet, but definitions recently adopted by the International Astronomical Union make that an incorrect use of the word "planet". If all these factors really are important for the evolution of life then we have to look for planets very like the Earth, accompanied by a satellite very like the Moon, revolving around a star very like the Sun, at much the same distance as we are from the Sun. Despite the advances of the last decade, we still have no idea how common such a combination may be.

Perhaps the major handicap in our search for life elsewhere in the universe is that we still do not know how life began. Many people believe that life will automatically emerge where conditions are right for it. We can make some intelligent guesses about what those conditions might be but we do not know for certain what they are. If we did, we could produce life in the laboratory. As mentioned above, it was thought at one time that life might have begun in tidal pools, or at least in the shallow waters near the shore, but the surprising discovery of abundant life in the deep ocean vents, in an environment that we would otherwise have thought to be toxic, has changed our ideas and there are some who champion the notion that life actually began in those vents. One theory that goes in and out of favour is *panspermia* – life originated elsewhere in the universe and was brought here in "seed" form, possibly by comets. The idea is associated with the name of the Swedish scientist, Svante Arrhenius[8], who wrote about it in the early twentieth century, but it is older and has been revived much more recently by Hoyle and Wickramasinghe[9], who also argued that viruses are still being showered on the Earth by comets, causing, among other things, epidemics of influenza. The theory of panspermia appeals to those who see the emergence of life as highly improbable, but who do not wish to invoke Divine intervention. Much more time and space are available if life could have originated anywhere in the universe. If the hypothesis is correct, the probability of finding life elsewhere in the universe is greatly increased, since there *must* be other life somewhere and, presumably, other suitable planets will have been showered with viruses or some form of DNA or RNA. Many "organic" molecules, some of which can be regarded as the building blocks of life, have been found in interstellar space – another environment that is expected to be hostile to life – and that favours the hypothesis of panspermia. (The word "organic" is used here in the technical sense of organic chemistry, namely, to refer to compounds of carbon other than its two oxides and the metallic carbonates.) While the dissemination of viruses or DNA *within* a galaxy is just about credible, however, their dissemination *between* galaxies is much less so. More importantly, from the point of view of understanding how life began, the hypothesis of panspermia simply pushes the problem one stage back into regions of time and space where we can no longer study it.

If and when an Earth-like planet is detected around a normal star, obviously it will receive a lot of attention directed to establishing whether or not life is present on it. Since no such planet has yet been detected, many people have opted for a different approach – studying the radio emissions from Sun-like stars in the hope of finding signals containing intelligible information. We ourselves, after all, have been sending first radio, and now television, signals into space for approximately a century, even though at first we did so unintentionally and, since those signals are not directed toward a specific target, they will be very weak by the time they reach any potential recipients there may be. Nevertheless, a civilization that may exist "out there" that tunes into the right frequencies will be able to receive our signals, although whether or not it will be able to make any sense of "reality TV" is another question! We, of course, do not know where to look, or what frequencies to use, but much data has been gathered and it is at least possible that some day we shall stumble on something.

The difficulty of the above approach is establishing meaningful two-way communication. Think of the time involved. Suppose we should find a signal coming from the direction of our nearest Sun-like neighbour, α Centauri. (In fact, this star is double, or triple if the slightly closer Proxima Centauri is part of the system, and, although a careful search has been made, no evidence for any planets around either component has yet been found.) Light and radio waves take over four years to travel from those stars to us, so to get an answer to a simple question we would have to wait over eight years. That is the most favourable case: it is much more likely that any contact would be with a more distant star and that we would have to wait a century or more for the two-way trip. If the other civilization is in another galaxy, the wait-time becomes prohibitive – millions of years – and the problems of generating sufficient signal-strength become much greater. Imagine asking for advice on how to avoid a nuclear holocaust or global warming, and having to wait over a million years for the answer!

On the other hand, perhaps the other civilization is putting out its equivalent of the *Encyclopædia Britannica*, or *Wikipedia*, containing all its collected wisdom. Once we have located the transmission, all we have to do is to receive it and to study it – but could we understand such a transmission? It is generally argued that a civilization able to broadcast its existence and knowledge in that way must have technology like our own and, therefore, a similar understanding of science, and this should help would-be code-breakers to decipher the message. Our representatives, on our part, gave much thought to how we should present ourselves in the two *Pioneer* spacecraft that were sent out of the solar system. It is extremely unlikely that those two craft will ever be intercepted, and I am far from certain that their meaning will be obvious if they are. After all, we know that other creatures on our own planet are able to communicate with their own kind, and we have barely begun to decipher their signals. The dance-language of bees was deciphered some time ago, but the songs of whales and the clicks of dolphins are still largely mysteries to us. Even our closest relatives, the chimpanzees, are not able to tell us very much. We have taught one or two of them a few of our words, and thus had very limited conversation with a few individuals, but no real exchange of "ideas" is possible, and may never be so, even if other creatures can be said to have "ideas". We all live on the same planet and so have some familiarity with each other's environment; if we were suddenly presented with intelligent communications from another planet, even without the complication of time-delay, could we expect to understand those signals any better than we do the signals of our fellow Earthlings? Even within our own species, the history of encounters between European colonizers and the aboriginal peoples they so often dispossessed does not predispose us to optimism on this score.

An even more important question is: how like us would any other intelligent beings be? Evolutionary biologists no longer speak with one voice on this matter. Stephen Jay Gould[10] and, before him, George Gaylord Simpson[11] argued forcefully that we could not expect to encounter extraterrestrial beings even remotely like ourselves. There were, they argued, too many contingent events in evolution – such as, we may now cite, the collision of an asteroid with the Earth that many people believe killed off the dinosaurs at just the right time for mammals to take over. Gould expressed the idea memorably by comparing evolution to a movie film that would be different every time it was rerun. More recently, Simon Conway Morris[12] has argued the opposite equally forcefully. He stresses the importance of the phenomenon of *convergence* in evolution. Convergence is the independent finding by unrelated organisms of similar solutions to the problems posed by the need for adaptation to the environment. Perhaps the most obvious example is the evolution of fins and fish-shaped bodies by mammals such as whales and dolphins that have returned to the sea. Morris, however, takes the camera eye (discussed in the last chapter). Contrary to the creationists, who claim that such a structure could not have arisen from chance mutations by natural selection, Morris asserts that it has so arisen several times, independently. There are so many examples of this kind of convergence that Morris believes that "rerunning" the movie of evolution would, in fact, produce much the same assortment of creatures that we have now. It is almost as if some body forms are more probable (in the Bayesian sense of prior probability) than others. Similar ideas have been expressed by Stuart Kauffman[13], who stresses that complex systems, even though they are of a kind that no-one would consider to be living, spontaneously order themselves in one of a limited number of ways, even though many others are theoretically possible. He believes that similar processes in living systems are comparable in importance for evolution to natural selection itself. From his point of view, the "arrival of the fittest" is as interesting as its survival – a turn of phrase that, according to George Johnson[14], goes back to the great Dutch botanist Hugo de Vries. Perhaps self-aware intelligent beings will always be about six-feet tall, walking on two legs, possessing forelimbs with opposable thumbs and eyes that face forward.

The arguments of the preceding paragraph refer, of course, to evolution here on Earth; but we have seen that, in all probability, any planet on which intelligent life can evolve will be closely similar to the Earth. Therefore, by Morris's argument, the optimum body patterns will be very similar to the ones we are familiar with already. Perhaps, after all, we will be able to understand the messages of intelligent beings if and when we intercept them.

We come back to the still unanswered question: how likely is it that there is intelligent extraterrestrial life and how likely is it, if there is such life, that we can contact it? Morris takes a pessimistic view: the conditions for beings like us to evolve are so special that they will have appeared only a relatively few times in the whole of this vast universe. That means we shall be so widely separated from our counterparts that meaningful contact with them is highly improbable. On the other side are people so convinced that there is a multitude of other civilizations out there that they are monitoring the sky every night, searching for messages from them.

Whatever our own opinions on these questions might be, we probably all agree that conclusive evidence that we are not the only intelligent, self-aware beings in the universe would greatly affect how we think about our own relationship to that universe. The nature of that relationship raises religious questions, even if the idea that the discovery of other intelligent life-forms within the universe would undermine religious belief has been shown to be an exaggeration. Many of us instinctively feel, along with Dean Inge, that so vast a universe can hardly have been created for our benefit alone. Strictly orthodox Christians may still want to insist that Christ's redeeming sacrifice was a unique event for the whole cosmos, just as strictly orthodox Muslims would insist on the uniqueness of the *Qur'an*, and it is the religions that claim uniqueness that are most likely to be troubled by the thought of the existence of extraterrestrial life. I see no reason why Hindus or Buddhists should feel their beliefs to be threatened in any way from that quarter. Even Muslims, perhaps, could accept that Allah had revealed the *Qur'an* or its equivalent to other planetary civilizations. If there is a problem at all, then, it is primarily one for Christians, but even they may find a way of reconciling their traditional claims of uniqueness with the possibility of the existence of other intelligent beings elsewhere in the universe. As we saw in the previous chapter, there are even those who argue that the existence of many universes is consistent with notions of Divine creativity. Still others may feel that such prolific creativity echoes Hindu belief in the dance of Shiva. If intelligent life evolves, on average, only once in a galaxy, or for that matter, once in a hundred galaxies, there would still be a lot of it in the universe – but very little chance of civilizations being in contact with each other. The cosmos could be teeming with life and we would never know about it. Perhaps, when one reflects on how different civilizations have interacted here on Earth, it is just as well.

References:

[1] Inge, W.R., *God and the Astronomers* (Longmans Green & Co., London, 1933), p. 249.

[2] Kepler, J., *Somnium, Sive Astronomia Lunaris*, originally published posthumously in 1634. An English translation by Patricia Kirkwood, with introduction and annotation by John Lear was published as *Kepler's Dream* by University of California Press, Berkeley and Los Angeles, 1965.

[3] Fontenelle, B. le Bovier, *Entretiens sur la pluralité des mondes*, 1686, Paris; English translation by H.A. Hargreaves with an introduction by N.R. Gelbart (University of California Press, Berkeley, 1990).

[4] Milne, E.A., *Modern Cosmology and the Christian Idea of God* (Oxford University Press, 1952), p. 153.

[5] Peters, T., *Astrotheology and the ETI Myth*, in *Theology and Science*, **7**, 3-30, 2009.

[6] Wells, H.G., *War of the Worlds* (William Heinemann, London, 1898).

[7] Waltham, D., *Astronomy and Geophysics*, **48**, pp. 3.22-3.24, 2007.

8. Arrhenius, S., *Worlds in the Making* (Harper and Bros., New York, 1908).
9. Hoyle, F. and Wickramasinghe, N.C., *Diseases from Space* (J.M. Dent and Sons Ltd, London, 1979).
10. Gould, S.J., *Wonderful Life: The Burgess Shale and the Nature of History* (W.W. Norton & Company, New York, 1989), pp. 287-290.
11. Simpson, G. G., *This View of Life: The World of an Evolutionist* (Harcourt, Brace and World Inc., New York, 1964), Chap. 13.
12. Morris, S.C., *Life's Solution* (Oxford University Press, 1995), esp. Chaps. 10-12.
13. Kauffman, S., *At Home in the Universe* (Oxford University Press, 1995).
14. Johnson, G., *Fire in the Mind* (A.A. Knopf, New York, 1996), p. 245.

CHAPTER 6
The Mind and the Brain

> We are such stuff
> As dreams are made on, and our little life
> Is rounded in a sleep.
>
> William Shakespeare, *The Tempest*

Whatever Shakespeare believed himself, and some have argued that he secretly remained loyal in Elizabethan England to the Roman Church, he often put into the mouths of his characters some surprisingly modern and atheistic statements. It is a small step from Prospero's summary of the human condition to that with which Francis Crick[1] opened his book, *The Astonishing Hypothesis*:

> The Astonishing Hypothesis is that "You", your joys and your sorrows, your memories and your ambitions, your sense of personal identity and free will, are in fact no more than the behavior of a vast assembly of nerve cells and their associated molecules. As Lewis Carroll's Alice might have phrased it: "You're nothing but a pack of neurons".

In the previous chapter we considered the possibilities of finding intelligent, self-aware life in other parts of the universe, without specifying very closely what we meant by intelligence and self-awareness. Those questions and related ones have been attracting a great deal of attention in the last several years. In one sense, progress in the field is at least as rapid as progress in astronomy and genetics, and may very well turn out to be even more important for our understanding of ourselves. Copernicus dethroned us from the centre of the universe, Darwin pointed out our kinship with the rest of the animal creation, and modern research on the brain challenges others besides professional philosophers to think about the very nature of our personhood. The new methods of medical imaging that are now available make it possible to study the brains of human beings while the subjects are acting or thinking, and we can discover which areas of the brain are active as we think about given topics, try to move, or try to formulate our thoughts in words, and so on. Although Susan Greenfield[2], in her book *The Private Life of the Brain*, maintains that the brain works as a whole, and it is misleading to think in terms of particular regions of the brain for particular thoughts and actions, this research on cerebral activity forces us all to face the question: is the mind simply a function of the brain? If we answer "no" to that question, another one arises: can we avoid the dualism of mind and matter? From this point of view, progress in mind-brain research is not so rapid; we are scarcely any further ahead than Hume and Descartes, or even Plato, were. We are still facing the same problems that they faced, and we have not yet reached a consensus about the different views they presented. The basic choice is still between monism and dualism: that is to say between belief that there is only one basic "stuff" (or "substance", to use the old term), or recognition of two basic stuffs: mind and matter. The monist position can be further subdivided into materialism which, as we have said, sees mind simply as a function of the brain, and mentalism (or idealism), which accords the priority to mind, rather than matter. While idealism has been important historically, especially in the work of Berkeley and Plato, the modern debate is largely one between materialism and dualism.

Gilbert Ryle[3], in his book *The Concept of Mind*, strongly criticized mind-body dualism, dismissing it as "the dogma of the Ghost in the Machine". Others have gone much further and regard the mind simply as a function of the brain. They believe that following the lines of research outlined in the previous paragraph will eventually tell us everything there is to know about the mind. Crick expressed this fairly general attitude (described by Peter Dodwell[4] in his book *Brave New Mind* as the "position of virtually every neuroscientist") in a memorable turn of phrase, and while he honestly admitted that he could not prove the "Astonishing Hypothesis", he believed in it himself and was confident that, sooner or later, its truth would be conclusively demonstrated. On this view, something similar to our minds or consciousnesses would appear in any sufficiently complex neural network, such as we could hope to reproduce as soon as we become sufficiently skilled at making silicon chips, or whatever replaces them. This point of view is argued strongly by Patricia Churchland[5] in her book *Neurophilosophy*, in which, in the context of Libet's experiments (see below), she states that she "could not find any grounds for concluding that mental states are distinct from brain states". Another philosopher, of a quite different persuasion, Richard Swinburne[6], argues forcefully for making just such a distinction, and in his book *The Evolution of the Soul* adopts a dualist position, seeing the mind (or soul) as distinct from the brain while admitting that, in our experience at least, the former can act only through the latter. Causation, Swinburne believes, can act both ways: brain states can cause mental states and vice versa. Roger Penrose[7], in his book *The Emperor's New Mind*, concluded that we would not be able to build a digital computer that would reproduce the human mind; he left open the question of whether or not a quantum computer – a concept in its infancy at the time that he wrote – could do so. More recently, John Leslie[8] in his book *Infinite Minds* has explored that possibility further, but our own ability to make quantum computers still lies some time in the future[9].

Peter Dodwell, in the book cited above (pp. 90-96), makes a very clear case for distinguishing between "cause" and "explanation" or "reason". Brain states may *cause* our conscious thoughts and feelings, but they do not *explain* how those states are translated into our feelings of joy or depression, or whatever else we may experience. Penrose and Dodwell are arguing against the reductionism of Churchland and Crick. The Astonishing Hypothesis is of course pure reductionism, as Crick, a card-carrying "rationalist", would cheerfully have admitted. He would have supposed that the entity we call Francis Crick ceased to exist, if not at the moment when his heart and breathing stopped, certainly no more than a few minutes later, when his brain would have suffered irreversible damage. The irony is that, if he was right, he will never know, whereas if he was wrong, he will certainly have found out by now!

Some of the things the mind does, while indicating a strong connection between it and the body, are rather hard to explain in reductionist terms. The placebo effect in medicine is a well-known example, which I will have occasion to discuss again in Chapter 8. If the mind is convinced that the body is being treated for some disease, the body often gets better even if the "treatment" is nothing but a sugar-coated pill with no medicinal value at all. The opposite effect, sometimes called "nocebo" is also known and has recently been discussed by Helen Pilcher[10]: people can become convinced that a neutral substance will cause them harm and will indeed show symptoms of illness as a result of ingesting it. Perhaps still more surprising are cases of imaginary pregnancy. A woman can wish so strongly to have a child that she may show all the signs of pregnancy: cessation of periods, morning sickness, distension of the abdomen, cravings for particular foods, and even "quickening" of the non-existent baby in her womb. This condition is less frequent, at least in the Western World, than it once was, presumably because there is no longer as much pressure on women to have babies as there used to be, but it does occur, and a fictionalized account of a real case is given by V.S. Ramachandran and S. Blakeslee[11] in their book *Phantoms of the Brain: Probing the Mysteries of the Human Mind.* The description of such a sequence of events as "nothing but a pack of neurons" is not immediately convincing.

The real problem with reductionism, however, is that it explains nothing. In the perennial dispute as to whether mind or matter is prior, it always seems to be assumed that mind is something rather elusive and mysterious, whereas matter is hard, inescapable fact. At the level of fundamental particles, however, the nature of matter is just as elusive as that of mind. The dual wave-particle nature of matter is by now well known, as is also the equivalence of matter and energy. Just because of the dual nature of fundamental particles, we have to be very careful in talking about their "sizes", but there is a sense in which we can say that much of an atom is empty space. The inability of one solid body to penetrate another without at least one of them being destroyed in the process arises from the electrostatic repulsion between the electrons in each of them, which also accounts for the sensation of hardness that our sense of touch gives us, as I suggested in Chapter 2. Eddington[12] once remarked of electrons that they are "something unknown... doing we don't know what." From that point of view, reducing all mental phenomena to chemical reactions or the firing of electrical synapses in the brain is simply replacing one mysterious phenomenon by others that, ultimately, are equally mysterious. Physicists, who work with inanimate matter, seem to be more aware that its ultimate nature eludes our understanding than are many of those who work in the biological sciences or who study the working of the brain. Ironically, members of those last two groups seem to be more likely to adopt the kind of old-fashioned materialism that many physicists have rejected. The most the materialists can claim is that, since they are monists rather than dualists, they have reduced the number of mysteries in the universe by one.

A generation or so before modern methods of brain research were available, two dualists, the philosopher K.R. Popper and the neurophysiologist J.C. Eccles[13] collaborated in writing the book *The Self and its Brain* (the book in which Popper introduced his three worlds, already discussed in Chapter 2). Eccles, who, exceptionally it seems among brain scientists, was not only a dualist but a theist, contributed a detailed account of the fascinating experiments of Benjamin Libet and his associates (Section 9.2, pp. 256ff.). According to one recent report by C. Frith[14], the significance of this work is still being debated by those working in the field. Most books on the mind-brain relation include some account of Libet's work, but they are not always completely consistent with each other, especially as regards Libet's own views of the significance of his work. Fortunately, not long before his death, Libet summarized his own work and his own beliefs about its interpretation in the book *Mind Time*[15]. Toward the end of the book (p. 216) he wrote, quite unequivocally, "I can say categorically that there is nothing in neuroscience or in modern physics that compels us to accept the theories of determinism and reductionism."

Because Libet began working before the modern methods of brain imaging were available, he and his colleagues collaborated with neurosurgeons to perform experiments (with patients' fully informed consent) on brains exposed for some necessary surgical procedure. Three important results were obtained: (i) when a series of weak electrical pulses at the threshold of detection is applied to a part of the cerebral cortex associated with a particular part of the body, subjects experience sensations at the appropriate places on their skins, but only if the pulses persist for at least half a second; (ii) applying such a series of pulses *after* a threshold stimulus has been applied to the appropriate part of the skin can mask the sensation of touch; (iii) there is detectable activity in the brain for about half a second *before* the subjects become consciously aware of decisions to perform simple acts, such as lifting a finger or flicking a wrist, which they believed they had made quite voluntarily at a time of their own choosing. This last result was not entirely new; an Austrian scientist, H.H. Kornhuber, and his associates, had earlier devised methods of measuring through the scalp tiny electrical currents generated in the brain and found that activity in the brain began up to 1.5 seconds before his subjects were conscious of any intention to move a limb. Libet's group repeated the experiments with more rigorous procedures, and confirmed the result, but reduced the amount of time involved. Still more recent work, cited by Martin Heisenberg[16], however, suggests that the interval between the brain initiating an action and our conscious awareness of intending that action may sometimes be even longer than Kornhuber first found. Nevertheless, Heisenberg argues for the validity of human free will.

The last of Libet's experiments has perhaps been the most controversial – it was certainly the one discussed in the conference report referred to above – because it might appear, prima facie, to be evidence that our minds are, indeed, no more than our brains and that we do not have free will. If activity begins in the brain *before* we are conscious of having made a decision to move a limb, how can that decision have been voluntary? Eccles, on the other hand, saw in the time delay support for his dualistic hypothesis that a self-conscious mind could read the brain and affect it; as he put it: "…the self-conscious mind exercises a superior and controlling role upon the neural events." It is precisely in the context of such diverse interpretations of Libet's results that Patricia Churchland made the remark quoted above. Libet's own considered interpretation (pp. 137-156) is that the decision to move the limb is not voluntary, but once we become aware of it we *can* veto it. He found experimental evidence for that statement, which, after all, is consistent with the everyday experience of each one of us. We can restrain ourselves from an intended action that we might even have begun to perform. Anyone of us, in a moment of anger at some offence, real or perceived, might feel the urge to lash out and hit someone. Fortunately, most of us refrain from that sort of action, almost as soon as we become aware of the desire to perform it. In that sense, Libet argues, we do have free will. Swinburne does not regard Libet's experiments as relevant to the question of free will because no belief in the worth of the actions on which the experiments are based is involved (see New Appendix G of the book already cited). He does, however, argue for free will on the basis of what he terms "human counter-suggestibility" (pp. 252-259), which seems rather similar to Libet's argument that we can veto our actions.

Libet makes a strong distinction between our *detection* of a stimulus and our *awareness* of it. Other experiments that he performed convinced him that subjects had detected threshold series of pulses that lasted less than half a second, even though they had not been aware of the pulses. We can actually begin to react to a stimulus before we become aware of it. When we do become aware, we think that we have detected the stimulus as soon as it was applied, and Libet supposed that, in some way, the brain can refer the cortical stimulation back in time, giving us the illusion of continuous and immediate experience. He pointed out that athletes or musicians who try consciously to think about the movements that they must make perform badly; good performers, in fact, are not aware of what they are doing while they are doing it.

How conscious awareness arises from the purely physical brain has come to be known to neuroscientists as the "hard problem" and this is precisely the problem that Dodwell has in mind when he insists that, although brain states may *cause* conscious events, they do not *explain* them. Libet (pp. 162-163) favours the view that consciousness is an emergent property of the brain that cannot be predicted from knowledge of brain states, however comprehensive we may eventually be able to make that knowledge. He stresses several times that, even if we could know the state of every neuron (an estimated 100 billion) in the brain of a given person, we could not deduce from that knowledge the content of that subject's inner experience and feelings. He also posits that each of us has associated with our brain a "conscious mental field", whereby different parts of the brain can affect each other, even if there is no direct nerve connection between them. Surgical severing of the connections between the two hemispheres of the brain is sometimes carried out to control epilepsy. People who have been subjected to this procedure continue to feel that they are still one person, even though, in many respects, the two hemispheres function separately. Libet also describes possible experiments to test his theory that could be carried out (with the patient's and the neurosurgeon's consent) on people who, for some reason, have to have part of their brain removed. He considers the question of whether or not introducing a conscious mental field is a form of dualism: it is not, in the Cartesian sense of proposing that there are two separate substances, but the field is an emergent property not directly observable by physical means. It is what Churchland calls "property dualism" and Libet admits that it is reintroducing the ghost in the machine; he remained agnostic on the possibility of any form of a personality's survival of the death of the body.

In the book by Ramachandran and Blakeslee cited above, reference is made to other experiments on brain activity by Michael Persinger[17], who showed that stimulating the temporal lobe of the brains of people can produce for them something very like religious experience – leading the materialists to dismiss such experiences as nothing but the creation of our own brains. The temporal lobe is the part of the brain implicated in one form of epilepsy and, apparently, epileptics are more likely to have religious experiences than are the rest of us. It has been suggested that St Paul was an epileptic and his fall from his horse on the road to Damascus was simply a dramatic seizure; or even that Mohammed was an epileptic. Many Moslems, of course find this latter suggestion offensive – perhaps because they fear that it would compromise the authenticity of the *Qur'an*. Religious believers need not take offence; the fact that religious experiences can be simulated in the laboratory does not necessarily imply that spontaneous ones are not genuine. The lives of both St Paul and Mohammed were transformed by their respective experiences in ways that are not obvious in the stories of those who have had simulated experiences. Sulphur Mountain, near the town of Banff in the Canadian Rockies, can be ascended by a fairly easy trail or by a gondola lift. I have made the ascent both ways: the view from the summit is the same, whichever way you choose, but the satisfaction of having made the climb makes it all the more rewarding for those who choose the harder way. I have little doubt that reaching a summit by arduous rock-climbing brings even greater rewards, but my fear of heights prevents me from finding that out for myself. I am not perturbed, therefore, by the thought that people can have a religious experience by putting on a cap that contains magnets that stimulate the temporal lobe. That no more invalidates the experiences of the great mystics than riding to a mountain summit in a cable car invalidates the experience of genuine mountaineers.

That being said, it is obvious that Persinger himself believes that he has demonstrated that the experience of God is an illusion, although he also believes that having this illusion must have had adaptive value in the early history of the human race. He writes that it is a fundamental principle of behavioural neuroscience that "all experiences are generated by or correlated with brain activity" but then appears to forget the second half of the alternative. According to his own account, he has not only simulated experiences of the

Divine in the laboratory, but also "out-of-the-body", "near-death" and "Third-Man" experiences. The last named, to which I alluded briefly in Chapter 3, has been felt by many explorers in dangerous situations, and has recently been described in a book by John Geiger[18], who also prefers an explanation for it in terms of a special kind of brain activity triggered by the danger in which the percipients find themselves. Of course, there must be activity in the brain "correlated with" all these experiences because it is through the brain that we become conscious of them, but, again, that does not necessarily imply that the experiences are "generated by" the brain activity. If one is wedded to materialism then, of course, one sees the brain activity as the cause and the experience as the effect. It is, however, at least possible that the roles are sometimes the other way about. Some of the people who have experienced the "Third Man" are convinced that they were saved by Divine intervention, or at least by a guardian angel. If you are convinced that discarnate entities cannot exist, you have to find explanations such as those Persinger and Geiger favour, but if you are willing to grant at least the possibility of such entities, then they could be the cause of the brain activity.

Similarly, materialists have to try to explain away the so-called "near-death" experiences because, if they are what many of those who have had them believe them to be, they completely undercut the materialist position. Thus, the experiences have been dismissed because, again, they closely resemble some that have been induced by taking drugs and, therefore, can perhaps be explained as the result of changes in the chemistry of the brain. More recently, it has been suggested that this "near-death state" is related to dreaming states and that, potentially, almost all of us can experience it. Once again, how we interpret the evidence is affected by what we already believe. Of course we should be cautious when interpreting near-death experiences; presumably the only way anyone can find out for certain whether or not they are veridical is to die – and then, whatever the truth of the matter, those of us left behind still will not know! The illustration of the mountain is applicable here, too. The fact that drugs or normal dreaming can produce similar experiences does not in itself invalidate the experiences of those who have been genuinely close to dying. Two features stand out in the reports of those who believe that they have had veridical near-death experiences: the subjects have often changed their lives afterwards, living less selfishly and showing more concern for others, and they say that they no longer fear death. These seem to be arguments for something more than brain chemistry having been at work and dismissing such experiences as the activity of a "pack of neurons" once again lacks conviction.

Recently (see, for example, *New Scientist*, 1 September, 2007[19] and 10 October, 2009[20]) reports have appeared of attempts to reproduce "out-of-the-body" experiences (closely related to near-death experiences) in the laboratory, which appear to have attracted more attention than Persinger's. Feelings similar to those reported by some people while under anaesthetic, of looking at their own bodies from outside, can apparently be produced by showing someone an image of his or her own back while prodding that person's chest and simultaneously prodding a second rod at the camera which is providing the image of the back. While it may be of interest that out-of-the-body experiences can be produced in this way, I fail to see that it has any relevance to the genuineness or otherwise of the spontaneous experience. I am tempted to dismiss the experiments as an affair of smoke and mirrors, but no smoke is mentioned in the published report! Coincidentally, Libet (pp. 219-220) proposed an experiment that perhaps could throw light on the spontaneous experience. If patients could be found to volunteer to undergo a cardiac arrest under controlled circumstances, various objects in the operating room would be covered until the patients were unconscious and then uncovered only for the very few minutes that could be allowed before the heart had to be started again. Then, if any of those patients could describe the objects in question, there would be clear evidence that they had indeed been out of the body. For the sake of Libet's reputation, I should make clear that he was describing only something that would be technically possible; as he dryly remarked, "…it is not likely that the experiment would be approved by an institution's committee for the protection of human patients."

Other animals have brains, but we do not know whether or not they have minds. Scientific opinion on this matter has changed quite dramatically in my own professional lifetime. Behaviourism has been an influential school of thought, even in human psychology; forty to fifty years ago it was the dominant theory of animal behaviour: animals were believed to respond to stimuli in predictable ways and all talk of animal thoughts and emotions was dismissed as unscientific. Pet owners who protested were told that they were projecting their own thoughts and emotions on to their pets, or, at most, it might be conceded that domesticated animals might learn to simulate human emotions. This point of view, which is not so very different from Descartes' seventeenth-century ideas, was notably championed by B.F. Skinner in the twentieth century. While this view is still held by some, it is no longer as dominant as it once was – see a recent discussion by Frans de Waal[21], who points out that Darwin himself believed in continuity between the mental abilities of animals and human beings. One of the biggest factors that has brought many people back to this particular point of view of Darwin's is probably the increased understanding of our closest relatives, the great apes, brought about by the work of Jane Goodall (chimpanzees), Dian Fossey (gorillas) and Biruté Galdikas (orangutans). (In general, I am sceptical of claims that science as a whole would have developed very differently if more women had been involved from the very beginning, but it does seem to me significant that women led the way in this particular area.) We know now, for instance, that chimpanzees can and do feel intense emotions and that it is as wrong, in normal circumstances, forcibly to separate a chimpanzee mother and baby as it usually is to separate their human counterparts. Very recently[22], a group of those apes has been photographed while, apparently, grieving for one of their number who had died. We also know that our ability to make and to use tools is not the unique distinction we once thought it was. Chimpanzees, at least, do likewise; quite recently their ability to fashion and to use spears lethally has been documented. Our new understanding of primate behaviour is surely a factor in the rise of the movement to recognize what are called animals' "rights"; there is even talk in some jurisdictions of making chimpanzees legal "persons".

A recent discussion of the question of animal minds can be found in the book, *Evolving God*[23], which I will discuss in more detail in the next chapter, because King's chief concern is the problem of the origin of religion. She finds a considerable gap in this respect between human beings and the apes, but the discontinuity between human beings and apes on the one hand, and all the other primates on the other, is even more marked. She believes apes to be capable of empathy, of giving meaning to their communications, of following the rules of their group, and of imagination. Her book will probably be controversial among primatologists, as she herself recognizes, and I shall leave criticism to those as knowledgeable in the field as she is herself, but her observations appear to me to provide strong additional support for ascribing some degree of conscious self-awareness to the apes. Questions remain of course. Some people believe the difference between human abilities and those of the apes to be so great as to amount to a difference in kind. A more basic question is: how widely within the animal kingdom are whatever abilities apes possess distributed? Many people who keep cats, dogs and horses will stoutly maintain that those creatures, too, have emotional lives and personalities, although I have yet to see such animals using tools. Many people would maintain that all mammals have some rudimentary mental and emotional life, but might express reservations about reptiles and fish – the cold-blooded creatures – yet these, too, have their champions.

The difference between mammals and fish is amusingly illustrated by Susan Greenfield (p. 12 in the book that I have already cited). She regards emotions, rather than thoughts, as basic to conscious awareness, and suggests that if a child's pet goldfish should suddenly die in the night the parents could quickly replace the dead fish with a living one and the child would probably never know that the original fish had died. Such a deception would be harder to carry off if the dead pet had been a hamster, and no parents would even think of trying it with a cat or a dog. The mammals are capable of emotions, and their particular emotional responses give them individuality, so that even a child would probably quickly detect that a replacement had been made.

If warm-bloodedness is a prerequisite for a mental and emotional life, then we also have to consider birds. We sometimes use the phrase "bird-brain" in a very disparaging way, but, as we have seen, migratory birds are capable of astonishing feats of navigation by means that we do not yet fully understand. Adult birds can certainly show signs of distress at the loss of young, or even of eggs, when a nest is destroyed. Members of the crow family are particularly intelligent; they can often be seen dropping nuts or other hard objects from a considerable height in order to break them open and to get at the food inside. That might be considered just one step short of tool-use, but they can also use sticks to reach otherwise inaccessible things, and at least one case of a crow fashioning a hook out of a piece of wire has been documented[24]. I can frequently watch a heron fishing from a beach near my home. The bird always fishes at slack water and obviously knows when fishing will not be productive. On that particular beach there is a considerable range in the water levels at which slack water occurs, especially at low tide, so it is not the water level that prompts the bird to fish; neither, of course, is it the time of day. The Darwinian explanation, no doubt, is that herons and tides evolved together and natural selection long ago picked out those herons with a bodily rhythm close to the rhythm of the tides; the birds fish at slack water by instinct. It is, nevertheless, a sobering reflection that those particular "bird-brains" know, even if only by instinct, something that most of us have to look up in a tide-table! Even the humble sparrow is not to be despised. I have often refilled an empty bird-feeder only to find that the sparrows emerge from hiding as soon as my back is turned. Not only can they remember that the feeder is a source of food, but they appear to have learned to associate my approaching the feeder with the appearance of a fresh supply. On the other hand, they seem unable to recognise that someone who takes the trouble to feed them is unlikely to do them harm; they still scatter if they are at the feeder when I approach – although I understand that, with patience, some people have even persuaded wild birds to take food from their hands.

In the previous chapter I referred to Simon Conway Morris's arguments about the importance of convergence in evolution. In another book which he has edited[25] and which came to my attention only when my own writing was well advanced, he and other authors continue the discussion of the importance of this phenomenon. The book is the result of a symposium held at the Vatican Observatory, and the theme running through it is that intelligence itself is a result of evolutionary convergence. Ants[26], the crow family[24], the cetaceans[27] (whales, dolphins and porpoises) are all cited as examples of creatures in which intelligence has evolved and there is even a chapter on the intelligence of plants[28]. Much depends, of course, on how intelligence is defined, but one definition offered is the ability to solve problems and to change behaviour as a result of doing so. Plant roots can, apparently, distinguish between other roots of the same plant and those of another plant of their own species. They can also explore and find patches of soil particularly rich in nutrients, leading the plant itself to direct more roots to the same area. Many of the authors conclude (there are also some critical of the hypothesis) that there is a goal to evolution, namely intelligence, and that many only distantly related organisms have reached, or at least approached, this goal by quite independent paths. Bird brains and mammalian brains have quite different structures, yet intelligence is found in both groups. If plants can really be called intelligent, then a brain is not necessary at all, at least for some types of intelligence. Of course, this is a minority view among evolutionary biologists; "orthodoxy" is still that evolution is an undirected process and has no goal. The group around Conway Morris, however, have some thought-provoking data and good arguments on their side.

Leaving aside the fascinating question of whether or not plants can justly be called intelligent, there seems no doubt that the brains of many animals, even quite tiny ones, are capable of much; but how far can their possessors be said to have minds? Our brains and theirs control the bodily functions, but we also associate our own brains with our ability to make moral choices, our powers of abstract thought, even our sense of humour and other emotions, despite everyday language that distinguishes between the "head" and the "heart", and with our sense of personal identity. We see that mammals, at least, have some emotional life, but we do not know how far they may be capable of the richness of experience that we ourselves know. Since none of us have been animals, that is something that we cannot know. In the West, at least, we have tended to assume that only human beings can make moral choices and, therefore, be held morally responsible (although there were exceptions to that point of view in the Middle Ages). The whole point of the story in *Genesis* that is usually called "The Fall", or at least of its Christian interpretation, is that prototypical human beings deliberately chose to become morally aware, but it now seems that chimpanzees can, also deliberately, set out to deceive the human beings with whom they associate, and rather enjoy the fun when they succeed. The Fall, if there was one, may have occurred at a somewhat earlier stage of evolution than the uncritical reader of *Genesis*[29] might suppose. How early in evolution, then, did "The Fall" occur? Presumably, no-one would ascribe a moral sense to earthworms, or even to the sparrows that we discussed above, but many owners of dogs and cats are quite sure that their pets can feel guilt. Certainly those animals can display behaviour that, in a human being, would be interpreted as evidence of a guilty conscience, but there might, of course, be other explanations for that behaviour. The animal might be dimly aware that it has displeased its owner, and be fearing punishment, without having any concept of right or wrong.

One possible distinction between human beings and animals was discussed recently by Michael Corballis[30], who suggests that the uniquely human characteristic is *recursion*: we can think about thinking and are conscious that we are conscious. Popper made a similar point when he wrote (p. 144 of *The Self and its Brain*) "…I conjecture that only a human being capable of speech can reflect on himself." Corballis also goes on to point out that we can use tools to make tools. As we have just seen, other species can make and use tools, but we appear to be the only one that can use tools to make more tools – although both chimpanzees and crows can use a shorter stick to retrieve a larger one that will enable them to reach the thing that they really want. Related to this recursion is our experience of the passage of time, which gives us the knowledge of the inevitability of our own deaths. While this point may still be conjectural, it seems unlikely that any other animals share quite the same experience of time that we have, or know that they will inevitably die. Again, Popper makes the same point: "It is only man who may consciously face death in his search for knowledge."

Presumably, scientific materialists are not greatly exercised about distinctions between human beings and the other animals. The more complex the brain, the greater they would expect its potential to be. Human brains are the most complex we know (with the possible exception of those of whales and dolphins) but they are not different in kind from other mammalian brains. Therefore, from the materialists' point of view, we should not be surprised that characteristics previously thought to be uniquely human are found elsewhere in the animal kingdom, at least in rudimentary form. Ideas of a mind separate from the brain (Ryle's "ghost in the machine"), a soul, if you wish to use the term, are hypotheses of which the materialist, like Laplace "has no need". All our human abilities are simply products of the greater complexity of our brains. This may sound persuasive, but when we think of what the human mind has produced in science, literature, art and music – all features of Popper's "World 3" – to say nothing of some of the puzzling mental phenomena described above, we realize that we are being asked to swallow a pretty big statement!

Even the fascinating empirical results of the last decade or two do not, however, unequivocally force us to adopt the position that the mind and the brain are identical, and we find people of considerable intellect on both sides of the debate. Perhaps that is not surprising, since the debate between dualists and monists is at least as old as philosophy. Despite present-day suspicions of dualism and Ryle's arguments against the ghost in the machine, many continue to favour dualistic theories. Popper and Eccles, respectively agnostic and theist, both described themselves as dualists. Their book was published after Ryle's but before much of the recent study of brain activity became possible; they knew only of the work of Kornhuber and Libet. Popper's arguments in particular, however, are at such a fundamental level that I do not think they are invalidated by any new results yet available; a view, as we have seen, accepted by Libet himself. The obvious, if picturesque, objection to the idea of the ghost in the machine is: where are the levers with which the ghost manipulates the machine? Eccles had an answer to that question: he believed that modules in the cerebral cortex (containing up to 10,000 neurons each) might provide the ports through which the self-conscious mind could interact with the brain. More recently, Roger Penrose[31] (in *The Large, the Small, and the Human Mind*) has proposed that coherent quantum fluctuations in structures called *microtubules*, found in all cells and therefore present in individual neurons, might give rise to the conscious mind.

Recently, Andy Clark[32] has argued that our minds are not simply identical with our brains but extend throughout our bodies and even to the things that we use. We are all used to speaking of the "memory" of a computer, for instance, which is, after all, an extension of our own memories – as was a card-index in the days before we had computers – and, to that extent at least, can be considered to be a part of our minds. Clark means something more than that, however, in that he thinks that such objects actually play a role in what he terms *cognition.* That his ideas are controversial is amply testified by the amount of space in his book given to answering his critics, but they are of interest as evidence that, even on a materialist model of the mind, which Clark explicitly says he does not wish to question, brain states and mental states are not necessarily simply identical. Thus, the empirical evidence does not yet, by itself, enable us to answer conclusively the first of the two questions posed at the beginning of this chapter: are the mind and the brain identical? Once again, of course, each of us will interpret what evidence there is in the light of what we already believe. Scientific materialists will claim, with some justification, that the simplicity of their monism makes it superior to the dualism that seems inherent in any argument that the mind is something more than the workings of the brain.

That brings us to the second question posed at the beginning of the chapter: if mind and brain are not identical, can we avoid any form of dualism?

Materialism is, in some ways, a nice simple doctrine that harmonizes well with much of our everyday experience, but, as we have already seen, the nature of matter is as elusive to our thinking as is the nature of mind. Part of the appeal of materialism is that it is monist and, as I remarked earlier, reduces the number of mysteries in the universe by one. The alternative kind of monism, mentalism or idealism, does not seem convincing to most modern thinkers, although the argument basic to this book, that there may be (transcendent) aspects of the world not revealed to us by our senses, could, if accepted, make that alternative more plausible. A third possibility, mentioned, but not explored in detail by A.G. Cairns-Smith[33], in his book *Evolving the Mind*, is *neutral monism*: mind and matter are different aspects of the same basic stuff. Germs of this idea can be found in the writings of Spinoza, Leibniz and Hume (see Popper, again in the book by him and Eccles, and L. Stubenberg[34] on the website of the *Stanford Encyclopaedia of Philosophy*). In modern times, the idea became particularly associated with William James who converted Bertrand Russell to it, at least for a while. Neutral monism has attracted me because it *is* monist, although as I have read more about it I have found it less satisfying. So far, all our efforts to understand either mind or matter have ultimately brought us up against mystery so dualist theories leave us with two mysteries. *Any* form of monism shares with materialism the advantage of reducing the number of mysteries by one.

Russell[35,36] developed his ideas of neutral monism in two books, mainly in *An Outline of Philosophy* but also in *The Analysis of Mind*. At the time that he was writing these, he was reacting to the then new ideas in physics introduced both by relativity theory and quantum theory. Although the latter, in particular, has developed much farther since that time (*An Outline of Philosophy* appeared in the same year that Heisenberg published his principle of uncertainty) Russell showed a very clear grasp of both theories as they were understood at that time. To his mind (and here he explicitly quoted Eddington) the new physics had completely ruled out the kind of materialism that was popular in the late nineteenth century, yet he did not want to adopt an idealist (or mentalist) philosophy, as Eddington had done. The notion that mind and matter were different aspects of the same neutral "stuff" therefore appealed to him. Influenced by relativity theory, he claimed that ultimate reality consisted only of "events" (he had, of course, his own careful definition of what constituted an event, the details of which need not concern us here). Some events have properties that we describe as material, and others have properties that we describe as mental; the distinction between the two is not necessarily always clear-cut. James had argued that "consciousness" is the name of a nonentity[37] – in the sense that it is not an entity but a function. He maintained that the thought and the object thought of were but different aspects of the same thing, and that mind and matter were different aspects of what he called "pure experience".

I believe, then, that a response along the following lines to the two questions posed at the beginning of the chapter is at least defensible: if one accepts that the transcendent is a possible part of the external world in which we live, then we are not compelled either to believe that the mind and the brain are identical or synonymous, or to adopt some form of dualism. At this stage, we can only imagine what the "neutral stuff" may be like. Eddington was convinced that we cannot know the true nature of the universe, only something of its structure; that seems to fit well with the notion of "stuff" – or substance if you will – which can appear to us as either mental or material.

Russell developed neutral monism in a way that entailed the denial of a continuing self (here, perhaps, the influence of Hume on his thinking, not to mention that of the Buddha, can be discerned, as well as that of William James), and, in particular, a denial that anything of the individual person could survive the death of the body. In this respect, then, Russell was a precursor of Crick. It is of interest that there seems to be some common ground not only between Crick's materialism and philosophers such as James and Hume, but even with the Buddha himself. I do not think that the Buddha would have gone along with Crick's "nothing but a pack of neurons", but he might well have found some points on which the two of them could agree. Not all forms of neutral monism, however, rule out the existence of a continuing consciousness. Indeed, according to Stubenberg, several philosophers (Popper among them) have criticized neutral monists for not being truly neutral and holding on to what is really an idealistic philosophy in disguise. Many people ignore neutral monism, or, at best, mention it in passing without going into details. Perhaps that is because almost all of us find it all but impossible not to make sharp distinctions between mind and matter (or body) and, so far, neutral monist theories do not seem fully satisfactory; but perhaps it may be possible to work out a more satisfactory version of it than Russell's was.

The question of the post-mortem survival of an individual consciousness is, of course, related to these discussions although distinct from them. Just because materialism implies that there can be no survival of the personality after death of the body, the concept of an immortal soul that can so survive seems, at first sight, inescapably dualist. Plato and Pythagoras, with their notions that the soul is imprisoned in the body and can, after death, transmigrate to other bodies, including animal bodies, certainly seem to have been dualist. Transmigration is, of course, a characteristic belief of Hindus and Buddhists (either of which may possibly have influenced Pythagoras) who would not accept without question my statement above that none of us have been animals, but neither would they be happy with being labelled "dualist".

The concept of transmigration, or reincarnation, does not fare very well in Western thought. Scientific materialism rules it out as impossible, and orthodox Christianity frowns on the notion, maintaining that our eternal fate depends on what we do in this one life that we live in this world. Even one well-substantiated example of reincarnation would require considerable modification to either view. The work of the late Ian Stevenson[38], summarized in his books, strongly indicates that some young children can remember events of a past life which can be verified. His methodology has been strongly criticized by, for example, Paul Edwards[39] but the body of data is large. No doubt, the interpretation will be argued as fiercely as the results of experiments claiming to detect paranormal phenomena. As I understand him, Stevenson never claimed that he had proved that reincarnations happen, although he obviously believed them to be the simplest explanation of what he had found. Because even one well-established example of reincarnation would force those who believe in the identity of mind and brain to revise their notions, the sheer quantity of Stevenson's data should at least give us pause. Similar problems are raised by cases of multiple personality, discussed in detail by William James[40] in 1890, and well attested many times since. I might add, parenthetically, that I am not particularly attracted to the idea of reincarnation myself. Although my experience of the changes and chances of this fleeting life has so far been a happy one, I am not at all sure that I would relish the prospect of facing it again; but an idea that is central to at least two major world religions should be considered seriously.

The traditional Christian view has been that only human beings have immortal souls, and each of us has a unique soul. The sharp distinction that implies between us and the rest of the animal creation, of course, runs counter to much of the argument above. Aristotle also made a sharp distinction between us and the animals in proposing three types of soul: vegetable, sentient and rational – the last, of course, being uniquely human. His view, reinforced by traditional Christianity after Aquinas, has prevailed in the West, while some Christians have also found congenial the Platonic dualism of the body as the prison of the soul. Nevertheless, primitive Christianity emphasized not so much the "immortality of the soul" as the "resurrection of the body" and the latter can be interpreted as a monist doctrine, as Peacocke[41] has argued. As David Jenkins, a controversial former Bishop of Durham, notoriously remarked resurrection is not just a matter of the resuscitation of bones. If it were, it would be easy to show up the whole idea as a tissue of nonsense. I do not know whether or not any atoms that were present in my body at my birth are still there (I doubt it) but certainly many that were have long since been shed in dead cells, while many others have been added as my body grew. One of the principal constituents of our bodies is water, and every drop we drink has already, on average, passed through several other human bodies. Both these considerations lead to the conclusion that many souls would have claims on the constituent atoms of many bodies. The doctrine of the resurrection of the body, surely, is rather a recognition that our bodies are important parts of our identities and that we know each other by our bodily appearances. It is an assertion, whether or not you choose to believe it, that we shall continue (or perhaps resume) existence as recognizable individuals in some form, after our present bodies have inevitably died. I think that that belief can at least be reconciled with philosophical monism, although Swinburne, in the book already cited, weaves it into his dualism. The soul, he argues, is not immortal but can function again after being inert, if the brain is revived. What the form of the resurrection bodies will be is, perhaps, something on which it is unwise to speculate, as St Paul said very clearly. Unfortunately, between our time and his, many Christians *have* speculated and left us with imagery that strikes most moderns as nonsensical, and not even very appealing. Who does want to spend eternity floating on a cloud and playing a harp? I, for one, would much rather spend it debating with Socrates, although perhaps even he could be a bit boring after several aeons!

References:

[1] Crick, F.C., *The Astonishing Hypothesis* (Simon and Schuster, London and New York, 1994), p. 3.
[2] Greenfield, S., *The Private Life of the Brain* (Penguin Books, London, 2000), Chap. 1.
[3] Ryle, G., *The Concept of Mind* (Hutchinson, London, 1949; Peregrine Reprint (Penguin Books, Harmondsworth) 1963 and subsequently), esp. Chap. 1.
[4] Dodwell, P.C., *Brave New Mind*, (Oxford University Press, 2000), p. 36.
[5] Churchland, P. S., *Neurophilosophy: Towards a Unified Science of the Mind-Brain* (MIT Press, Cambridge, Mass., 1986), p. 486.
[6] Swinburne, R., *The Evolution of the Soul* (Clarendon Press, Oxford, revised edn., 1997).
[7] Penrose, R., *The Emperor's New Mind* (Oxford University Press, 1989), Chap. 9.
[8] Leslie, J., *Infinite Minds* (Oxford University Press, 2001), pp. 85-92.

9. Knill, E., *Quantum Computing*, *Nature*, **463**, 441-443, 2010.
10. Pilcher, H., in *New Scientist*, 16 May, 2009, pp.30-33.
11. Ramachandran, V.S. and Blakeslee, S., *Phantoms in the Brain* (Quill William Morrow, New York, 1998), Chap. 11.
12. Eddington, A.S., *The Nature of the Physical World* (Cambridge University Press, 1928), p. 291.
13. Popper, K. R. and Eccles, J.C., *The Self and its Brain* (Springer Verlag, Berlin, 1977).
14. Frith, C., in *New Scientist*, 11 August, 2007, pp. 46-7.
15. Libet, B., *Mind Time* (Harvard University Press, 2004).
16. Heisenberg, M., in *Nature*, **459**, 164-165, 2009.
17. Persinger, M.A., *The Temporal Lobe: The Biological Basis of the God Experience*, and *Experimental Simulation of the God Experience: Implications for Religious Beliefs and the Future of the Human Species*, in *NeuroTheology: Brain Science, Spirituality, Religious Experience*, R. Joseph (ed.), (University Press, San Jose, Calif., 2002), pp. 273-278 and 279-292. The first was originally published in *The Neuropsychological Bases of God Belief,* (Praeger, Santa Barbara, 1987).
18. Geiger, J., *The Third Man Factor*, (The Penguin Group (Canada), Toronto, 2009), Chap. 13.
19. Anonymous report in *New Scientist*, 1 September 2007, p. 20.
20. Ananthaswamy, A., in *New Scientist*, 10 October, 2009, pp. 35-36.
21. de Waal, F.B.M., in *Nature*, **460**, 175, 2009.
22. Sample, I., *The Guardian Weekly*, 7th May, 2010, pp. 32-33; see also <http://bit.ly/chimpstudy>.
23. King, B.J., *Evolving God* (Doubleday, New York, 2007).
24. Clayton, N.S., and Emery, N.J., see reference 25 this page, Chap. 7.
25. Conway Morris, S., (ed.), *The Deep Structure of Biology: Is Convergence Sufficiently Ubiquitous to Give a Directional Signal?* (Templeton Foundation Press, West Conshohocken, Pennsylvania, 2008).
26. Franks, N.R., see reference 25 this page, Chap. 6.

27   Whitehead, H., see reference 25 previous page, Chap. 8.
28   Trewavas, A., see reference 25 previous page, Chap. 5.
29   *Genesis*, Chap. 3.
30   Corballis, M.C., *American Scientist*, May-June, 2007, Vol. 95, pp. 240-248.
31   Penrose, R. (with critical comments by others), *The Large, the Small, and the Human Mind* (Cambridge University Press, 1997).
32   Clark, A., *Supersizing the Mind: Embodiment, Action, and Cognitive Extension* (Oxford University Press, 2008).
33   Cairns-Smith, A.G., *Evolving the Mind* (Cambridge University Press, 1996), p. 231.
34   Stubenberg, L., *Neutral Monism*, in *Stanford Encyclopaedia of Philosophy*, 2005. See website <http://plato.stanford.edu/entries/neutral-monism/>.
35   Russell, B., *An Outline of Philosophy* (George Allen and Unwin Ltd, London, 1927), esp. pp. 214*ff*.
36   Russell, B., *The Analysis of Mind* (George Allen and Unwin Ltd, London, 1921).
37   James, W., *Does "Consciousness" Exist?*, Journal of Philosophy, Psychology, and Scientific Methods, **1**, 477-491, 1904.
38   Stevenson, I, *Twenty Cases Suggestive of Reincarnation* (2nd Rev. Edn), (University Press of Virginia, Charlottesville, 1974); *Reincarnation and Biology: A Contribution to the Etiology of Birthmarks and Birth Defects*, (2 Vols.) (Praeger, Westport CT, 1997); and *Where Reincarnation and Biology Intersect* (Praeger, Westport CT, 1997).
39   Edwards, P., *Reincarnation : A Critical Examination* (Prometheus Books, Amherst, NY, 1996), Chap. 16.
40   James, W., *Principles of Psychology* (2 Vols), 1890 (Dover reprint 1950), Vol. 1, Chapter X.
41   Peacocke, A.R., *Theology for a Scientific Age*, enlarged edn. (SCM Press, London, 1993), pp. 140, 160-163.

CHAPTER 7
Reason and Revelation

> Earth's crammed with heaven,
> And every common bush is afire with God;
> But only he sees, who takes off his shoes,
> The rest sit round it and pluck blackberries…
>
> Elizabeth Barrett Browning, *Aurora Leigh*

Anthropologists tell us that every human society has some form of religious belief and archaeologists have found evidence, such as the cave paintings of Lascaux, for example, that that was so as far back as we can trace the existence of human societies. Neanderthals, too, buried at least some of their dead, which perhaps indicates that they had at least a dawning religious consciousness, and, as we have seen, Barbara King[1], in her book *Evolving God*, argues that some of the characteristics on which religion is built in human societies – particularly a sense of belonging – are already to be found in communities of the great apes. The antiquity and ubiquity of religious beliefs, then, are beyond dispute. To the believer, these very facts are indications that religious beliefs contain at least some truth; to the unbeliever, they are embarrassments that have to be explained away somehow, or perhaps, they argue, the very multiplicity of religions is itself evidence that *no* religion can be true. Some believers, at least, see the content of religious belief as having been revealed to us; the unbeliever, who does not accept that there is anyone (or anything) to do the revealing, will try to explain religion as a natural phenomenon, which reasonable modern human beings will cast aside as illusion.

We have seen that St Thomas Aquinas thought of reason and revelation as complementary. Reason served to show that the idea of God was at least not nonsensical and, he also believed, could demonstrate that God must be eternal, unchanging and immaterial, but we could come to know such things as God is love and the relationship between the persons of the Trinity only through revelation. That is a middle way that is now difficult to defend because modern sceptics and biblical literalists agree in seeing reason and revelation as antithetical. The materialist is committed to the belief that knowledge can be obtained only by applying our reason to the evidence of our senses, so we must discover everything for ourselves. As early as 1927, Julian Huxley[2] first published a book (of which, no doubt, his grandfather would have approved) with the title *Religion without Revelation*, which apparently found a very receptive readership and was revised and reprinted in the 1950s; many modern people dismiss the whole idea of revelation *a priori*. On the other side, literalists see the whole Bible (or other scripture) as the inerrant word of God which trumps the results of applying our reason to our sense-data whenever the two appear to conflict. Unfortunately, the one book in the Christian Bible that is explicitly called a revelation is full of rich imagery which, taken literally, seems utterly fantastic to most modern minds, but which significant numbers of Christians persist in regarding as literal prophecies of how the world will end, thus further discrediting the whole idea of revelation in the minds of many modern thinkers.

In this book, I have defended the position that there is a transcendent realm, not accessible to us through our normal sensory channels. If that position is granted, it is not logically impossible, however improbable it may seem to some, that there is some other means of making us aware of that realm, a means that we may call "revelation". Part of the problem with the concept of revelation arises from a misunderstanding of what it is. Not only sceptics but also some believers (such as those mentioned above) think of revelation in terms of the revealing of facts or, at least, propositions. Thus, biblical literalists think that the age of the Earth and the special creation of human beings are revealed facts – and in so thinking confirm scientific materialists in their low opinion of the very idea of revelation. The practice of the Roman Catholic Church also gives grounds for thinking of revelation in such terms. As recently as 1950, Pope Pius XII proclaimed that the assumption directly into Heaven of the body of the Virgin Mary, after her physical death, was a divinely revealed dogma – that is, something that the faithful must believe in order to be saved. On the other hand, natural theology, which is often contrasted with revealed theology, has always found a place in the thinking of the Roman Catholic Church. The twentieth-century critics of natural theology were mostly Protestants. I referred in Chapter 3 to James Barr's defence of natural theology[3], in which he points out that the boundary between the two kinds of theology is only vaguely defined. Those who believe that human beings can come to some knowledge of God through the study of the natural world, of necessity also believe that God is revealing Himself *through* that world and our studies of it. Some form of revelation, therefore, does seem to inhere in any kind of religion.

I wish to argue for a concept of revelation that may not be offensive to those who put their faith in the power of human reason. Just as I argued in Chapter 3 that faith was not primarily a matter of believing propositions, so here I argue that neither is revelation primarily the revealing of propositions or facts: biblical accounts of revelation have little to do with either. Consider, for example, the story of Moses and the burning bush, to which Elizabeth Barrett Browning alludes in the quotation at the head of this chapter. Although not much imagination is needed, when the rays of the setting or rising sun catch the surrounding vegetation, to see every common bush afire with God, too much emphasis on possible naturalistic explanations misses the whole point of the story. Even whether or not the story recounts an historical incident is less important than what it tells us about the nature of revelation. In its own terms, the story makes clear that God wasted no time revealing facts or propositions to Moses; even when Moses asked God's name, the answer was somewhat ambiguous. God had a job for Moses to do and told him in short order to get on with it, and something happened that changed the whole course of Moses's life – just as we saw, in the last chapter, the lives of St Paul and Mohammed were changed by experiences that they certainly thought of as revelatory.

Recently it has become fashionable to use the term "epiphany" for life-changing experiences. That word was simply the name of one of the major feasts of the Church, and until recently was not much known outside the regular Church membership, but since it means "showing forth" it is not very different in meaning from "revelation", and is not being misused – although perhaps sometimes trivialized – by modern fashion. The point of so many biblical stories of revelation is that the recipients are, indeed, led to change their lives, and this is an important truth whether or not the stories are historically accurate, or even whether or not the central characters even existed. The vision of Isaiah in the temple certainly changed his life, as did the much later dramatic conversion of St Paul on the road to Damascus. In other stories, the individual seems to have worked his own way through to a new understanding, which might be called a revelation: Hosea[4], for example, coming to see that God loved unfaithful Israel in much the same way as he himself loved his unfaithful wife. Again, the unknown person we call "Second Isaiah" came to see that God could act in history in the most unlikely way, through people like the one we call the suffering servant and whom Christians have identified with Jesus Christ Himself.

None of these stories concerns the revelation of facts or of propositions about God that we are expected to believe without further evidence, but all give us some insight into how people who have wrestled with the notion of the transcendent have come to view the nature of God and have found their own lives changed by their experience. One of the most dramatic stories in the whole Hebrew Bible is the sacrifice of Isaac[5] – averted at the last minute by Divine intervention – for which Richard Dawkins[6], in his book *The God Delusion*, saves particular ire. Despite Kierkegaard's[7] impassioned defence of Abraham in *Fear and Trembling* many would agree with Dawkins in feeling that the story as told reflects no great credit on either Abraham or God. Indeed, if this is an historical account, I myself would join with Dawkins in saying that the god it describes is not worthy of human worship. There is another way of looking at the story, however. The Bible contains many different, and sometimes inconsistent, ideas about the nature of God, being based on the writings of many different people at widely different times in history, and some of the authors were not above using drama, or even irony, to make their point. The Israelites were near neighbours of the Phoenicians, whose god, Moloch, *did* require the sacrifice of children, and was worshipped under other names by other neighbours of Israel. Indeed, passages in the books we know as the First and Second Books of Kings suggest that, even as late as the monarchical period, some Israelites worshipped Moloch, under the name Milcom[8]. The author of the *Genesis* story of Isaac may well have had his own revelation that the God of the Hebrews was not that sort of god, and did not require human sacrifice at all, and he may have been writing a polemic intended to bring his (or, perhaps, her) readers to that same conclusion, using all the narrative skill he or she could muster. From that point of view, the story is a huge success, as the wrath of Dawkins amply testifies. That wrathful reaction may have been precisely what our unknown author hoped for.

For the Hebrews, God revealed himself not by reeling off propositions they were supposed to believe, but by His mighty acts – principally, of course, the Exodus from Egypt, and later and secondarily, in the return from Exile. We can hardly worship a god who would inflict so much suffering on (mostly) innocent Egyptians as is described in the early chapters of Exodus. The story has many internal inconsistencies and there may never have been the sort of mass exodus of slaves from Egypt that it describes, since no account from the Egyptian point of view is known. Nevertheless, the story has served as a sort of Platonic "noble myth" for the Jewish people during the millennia in which they had no homeland (and still does so serve). Christians have seen it as a type of what they regard as a much greater deliverance wrought by the life, death and resurrection of Jesus Christ, who, at least as portrayed in the synoptic Gospels, did not reveal much in the way of facts or propositions. He spent most of his time healing people and telling stories, leaving His hearers to work out for themselves what those stories meant – a method of teaching still used by many of our indigenous North Americans. I shall discuss apparently miraculous healings in more detail in the next chapter, but whatever you believe about the healing stories of the Gospels, the fact that they are told reveals what the earliest Christians thought about the nature of God. The real revelation of the Gospels, however, is in the manner of Jesus's life and, above all, in the way He met His death. I doubt if He cares much about the minutiae of theological argument or the details of religious practice imposed by many churches on their members.

Etymologically, "revelation" and "discovery" are rather similar in meaning. Both words signify that something that had been hidden has become known. When we talk of discovery, we tend to think of ourselves, or at least some representative human being, as the agent, actively uncovering something previously unknown. When we talk of revelation, on the other hand, we tend to think of human beings as the passive spectators of the unveiling performed by a Divine agent. That seems to be true of the visions experienced by Isaiah[9] and St John the Divine[10], but Hosea and our hypothetical author of *Genesis* appear to have worked their own way through to their conclusions in a process that may have been very like discovery.

Numerous stories illustrate that scientific discovery is also not an entirely rational process. Like the biblical stories we have been considering, many of them may well be apocryphal, but they have survived because, again like the biblical stories, they illustrate a truth that is independent of whether or not they are historically accurate. Archimedes jumping out of his bath as he realized the solution to the problem of deciding whether or not the king's crown was made of gold; Newton sitting under his apple tree; Kekulé dreaming of the structure of benzene; and, one which is well authenticated, the French mathematical astronomer Henri Poincaré[11] suddenly seeing, as he stepped onto a bus, the solution to a problem that had vexed him for some time – all these are stories that show that the actual discovery comes in a flash of insight – a *eureka moment* or an *epiphany* in the popular sense of that term – rather than as a direct result of the discoverer's conscious thought, however important that might have been in the time leading up to the discovery.

In both discovery and revelation, the reasoning mind plays very little part in the actual event, but it prepares the ground for the event beforehand, and helps us to make sense of the event afterwards. Newton knew Galileo's results that showed that the distance travelled by a falling body was proportional to the square of the time elapsed during the fall, and he knew Kepler's laws of planetary motion. His flash of insight, whether or not it was triggered by the sight of a falling apple, was to see that both these results could be explained by the inverse-square law of gravitational attraction, and his genius consisted in being able to work out mathematically a proof of the insight that would convince other people. The insight itself, however, was neither more nor less rational than the vision of Isaiah in the Temple of Jerusalem. Indeed, Newton, whose motivations are now known to have been primarily religious, would probably have agreed that revelation and discovery are very similar processes, both requiring some measure of belief.

The whole burden of Michael Polanyi's[12] book *Personal Knowledge*, to which I have already referred, is that we can know, even in science, only by being committed to a belief which, nevertheless, we recognize might be mistaken. He writes (and the italics are his): "…*truth is something that can be thought of only by believing it.*" And, again, "…to avoid believing one must stop thinking." Eddington's insistence that seeking is the true spirit of both science and religion also fits well with this parallelism between discovery and revelation. Many people, however, see religion simply as a matter of dogma that the believer must accept, and think that seeking is characteristic only of scientific investigation. In Chapter 2 I suggested that scientific discovery and artistic creation had much in common; now I am suggesting that revelation has elements in common with both of them. That should not surprise us. Even on the human level an artistic creation reveals to us something about the artist so, if the universe is indeed a creation, its very existence reveals something of the nature of the Creator. That is the foundation of the old idea of two books and the whole concept of natural theology.

The middle ground on which we can defend revelation, therefore, is that it is not the revelation of theological propositions about God, still less of scientific facts about the natural world, but a life-changing experience that in itself reveals something of the nature of the transcendent. Although I am defending in this book the position that there are rational grounds for religious belief, there is a non-rational (or perhaps supra-rational) element in the concept of revelation, but we see the same sort of non-rationality in artistic creation and scientific discovery. I am deliberately avoiding the word "irrational"; the sceptic who wishes to apply it to revelation must reckon with the fact that it applies equally to the process of scientific insight.

Of course, those who reject all religious belief as delusion will want to search for a naturalistic explanation of the existence of such beliefs in all societies. An explanation popular among sceptically inclined scientists, which owes much to Freud's[13] discussion in *The Future of an Illusion*, is that gods and spirits, so far from revealing themselves to us, were human inventions, devised to explain the workings of the natural world by primitive peoples who had no scientific understanding. As their knowledge of the natural world increased, so the story goes, fewer gods were needed and the interventions of those remaining also decreased in number. Eventually, we were left with belief in one God Who hardly ever intervenes, and even that God is bound to disappear when we have the much sought after "theory of everything". Plausible though this schema may seem, it is invalidated by the fact that the major monotheistic religions appeared long before our ancestors had any detailed scientific knowledge of the natural world.

A somewhat different approach, which still sees religious belief as a perfectly natural phenomenon, is to argue that it confers an evolutionary advantage on the groups holding them. This is an application of E. O. Wilson's sociobiology, now more frequently termed evolutionary psychology, and the case for it has been strongly argued in the book *Darwin's Cathedral* by David

Sloan Wilson[14]. I admit that I approached D. S. Wilson's book in a somewhat sceptical frame of mind, thinking that a much simpler explanation of the presence of religion in virtually all human societies is that all of them contain some truth – and scientists are always supposed to prefer the simplest explanation when more than one is available! In fact, I found the book quite persuasive, partly because of the author's moderate and reasonable tone. D.S. Wilson stresses the benefits that a group (or society) receives from shared religious belief. Historically, the idea that natural selection can work at the level of groups as well as of individuals has been controversial among evolutionary biologists; many believe that natural selection acts only on individuals, and some even insist that it acts only on individual genes (the "selfish gene" of Richard Dawkins). Wilson is as concerned to win over those who still have reservations about the concept of group selection as those who might resist his claim that religion exists because it is adaptive. (Recently, E.O. Wilson and D.S. Wilson have jointly written an article defending the concept of group selection[15].)

Perhaps the most convincing of D.S. Wilson's case studies is of a complex of temples and villages in a mountainous area of Bali, and the irrigation of the villagers' rice paddies (see pp. 126-133 of the book cited above). The main temple is on the summit of a volcano, from which descends an elaborate system of tunnels and aqueducts. There are other temples and shrines at key points in the irrigation system and also in the villages and even individual houses. This complex of building and engineering works is the result of the collective effort of a large number of people from many different villages, and the cooperation is organized by religious authorities rather than secular ones. The religion is syncretistic, combining elements of an indigenous religion with those of many imported ones, and there are many deities. Nevertheless, the priests use their authority for the very practical purposes of regulating the water supply to the rice paddies, and controlling the spread of pests from one community to another. Religious control, in fact, provides a way of avoiding what has become known as the "tragedy of the commons". Maybe no individual and no village get all the water they want, especially in dry years, but all get some, and each village, as well as the whole complex, benefits by sharing. Religion is seen here as playing a very positive role to the benefit of the whole community as well as to that of the individuals that the community comprises, but that says nothing about its truth-content. This kind of justification for religious beliefs and practices commends itself neither to thoroughgoing materialists, who are reluctant to concede that religion can confer *any* benefits, nor to students of religion, who find the concept of evolutionary origins for religion unsatisfactory. The point I wish to make here, however, is that even if Wilson's hypothesis should turn out to be correct that does not rule out the possibility that most religions, if not all, contain some truth. We are not confronted with a situation in which we have to choose between rival explanations; each could be correct as far as it goes, without invalidating the other.

Mikael Stenmark[16], in his recent book, *How to Relate Science and Religion: A Multidimensional Model*, regards the sociobiological approach to the origin of religion as an example of what he calls "scientific expansionism" – an attempt by scientists to move into the sphere of religion and even to explain the very existence of religion in scientific terms. (He is equally critical of "religious expansionism" – the attempt by religious believers to dictate which scientific theories are acceptable.) Regardless of whether Wilson's arguments eventually stand or fall, I do not see that religious believers need to resist them, because naturalistic explanations of religious beliefs do not necessarily rule out the possibility that those beliefs are true.

Other authors also argue that religion has adaptive value; for example, Brian Hayden[17] in his book *Shamans, Sorcerers and Saints*, remarks that whenever people spend large amounts of time, effort and resources on a widespread but specific type of behaviour that persists for long periods of time, that behaviour is likely to have some adaptive benefits. The cave paintings in Lascaux, temples in what is now southern Turkey, and the well-known (but much later) Stonehenge and the pyramids of Egypt are all examples of projects that required considerable communal efforts over long periods. The earliest of these projects were undertaken when our ancestors were living barely above subsistence level, and must have required almost all the spare time, energy and resources available to the communities concerned. The explanation favoured by Hayden is again that this kind of collaboration, in the name of something greater than individual human beings, helped to strengthen the cohesion of the group, and that those who lived in groups and cooperated had a greater advantage in the evolutionary struggle and so survived.

The inculcation of a sense of community among believers is a feature of all religions. Wilson, Hayden and King in their respective books all stress the psychological importance of a sense of belonging to a group, which is of value both to the individual and the group. Wilson uses the importance of that sense of belonging to strengthen his argument that religious beliefs confer an evolutionary advantage. It is probably premature to venture a final judgement on such ideas which are, I suspect, as likely to be attacked by some biologists as by religious believers. Indeed, Richard Dawkins is clearly sceptical, perhaps partly because of his scientific reservations about group selection and partly because he is reluctant to concede that religious beliefs *could* be adaptive. Dawkins wants to argue instead that religion is a by-product of evolution. He also goes on to argue that morality is a by-product of evolution. It seems rather odd to argue that two of the things that most clearly distinguish human beings from other animals should be only by-products of evolution.

Hayden (p. 15) also reports that in many of the still surviving hunter-gatherer societies about ten per cent of the members are self-professed atheists or agnostics. Since the religions of such societies are thought to be similar to those of the earliest societies (that have not survived into modern times) this indicates that religious scepticism is as old as religion itself. There is other evidence for this statement. According to the poet (or poets) who wrote the Psalms, the fool said in his heart "there is no God"[18]. Or again, in another psalm, "Tush, they say, how should God hear it? Is there knowledge in the Most High?"[19] I do not necessarily wish to apply the Psalmist's assessment to our modern atheists, but these two psalms are evidence that there were unbelievers in antiquity – probably even before the time of most of the pre-Socratic philosophers. Among those latter, we find Xenophanes, whose remarks to the effect that, if horses and oxen could draw, they would represent the Deity as one of themselves, are often quoted. His own brand of monotheism is less frequently recalled, however: his target was not theistic belief, but anthropomorphism. Or, again, in the *Rig Veda*, believed to date to about 1000 BCE, we read:

> Whence has this creation arisen – perhaps it formed itself, or perhaps it did not – the one who looks down on it in the highest heaven, only he knows – or perhaps he does not know.[20]

We tend to think of scepticism as a product of the Enlightenment, but these examples show that it has coexisted with belief at least as long as there has been writing. The change wrought by the Enlightenment, therefore, was not so much the introduction of the possibility of religious scepticism, as the creation, at least in modern Europe, of a climate of opinion, in which individuals could be more open about what they did or did not believe. This provides another reason for dismissing the simple account of evolution from animism to monotheism, and eventually atheism, that I summarized above.

Another possible explanation for the ubiquity of religion has recently been discussed by Michael Brooks[21] in the pages of *New Scientist* in an article citing several investigators. The explanation is quite simply that human beings are predisposed to religious belief – in our modern computer-inspired jargon, our brains are "hard-wired" for belief in the supernatural. Once again, such a theory of origin does not rule out the possibility that religious beliefs are in some sense true, as is acknowledged by the researchers putting forward this particular hypothesis. Indeed, I would incline to the opinion that if our brains are indeed hard-wired for religious belief, then that is a very good argument for supposing that such beliefs contain at least some truth.

I introduced this chapter by seeking for explanations of the antiquity and the ubiquity of religious belief in human society. We have explored three: that all religions contain some truth and reveal something of the transcendent, at least to their own adherents; that religious belief has evolved in human societies because it offers an advantage to the group holding it; or that the human brain is so constructed as to predispose us toward some form of religious belief. Which one is most likely to appeal to us depends very much on the prior assumptions that we make. This is one more example of Bayes' theorem in action, even if the assigning of quantitative values to the various probabilities eludes us. There is no dispute that religious belief is extremely widespread, or even universal, in human societies, but we may interpret this fact in different ways. For religious believers, the observed fact is not unexpected since, for them, religious beliefs are telling us something that is true about the world; so it is no surprise to them that human beings began to speculate about that truth as soon as they could speculate at all. In Bayesian terms, believers estimate that the probability of the evidence, given their hypothesis, is high. For scientific materialists, however, convinced that religious belief is delusion, the probability of the evidence given *their* hypothesis is low. Therefore, although they may allow that early human beings, who had yet to learn even the most elementary truths of natural science, could be excused their delusions, they find it difficult to understand why modern human beings persist in them.

Of course, the existence of so many religions, especially of so many that claim to be *the* true religion, is used by sceptics as an argument for supposing that none can be true; but it is not very difficult for believers to modify their arguments to take this into account. All human societies can be supposed to have received some intimation of the Divine Nature, but some have received more complete revelations than others. It is no surprise that the believer's own religion is seen as the fullest such revelation. Unfortunately, the believer may go so far as to aver that his or her own religion enshrines the complete truth and that every other religion, or even another version of the same religion, is imperfect at best and, at worst, downright false. That is special pleading, of course, and there is no doubt that the conviction that one's own set of beliefs is the true faith has done much, perhaps more than any other single consideration, to discredit religion in the eyes of thinking people who might otherwise be sympathetic to it. Unbelievers see that conviction as the root of all the wars of religion that have plagued human history; for example: the Thirty Years' War, the long period of unrest in Ireland, the bitter rivalry between Moslems and Hindus in India, and, not least, the global conflict we are witnessing today between some Moslems and some Christians, with those on both sides claiming that they are engaged in a clash of civilizations, cultures and religions, and each convinced that God is with *them*. Religious intolerance certainly has a lot to answer for, but not all believers are intolerant of other beliefs and the final judgement on religion should not be based only on its worst aspects. Archaeology, anthropology and evolutionary psychology can hardly hope to help us decide which of our existing religions, if any, is the "true" religion, although they might help us to weed out one that is completely false: if a religion proved to be totally destructive of society, and therefore not adaptive, we could perhaps reject it as quite wrong-headed. Contrary to the opinions expressed by some other recent writers, however, I do not believe that that can be said of any existing major world religion.

References:

[1] King, B. J., *Evolving God: A Provocative View on the Origins of Religion* (Doubleday, New York, 2007), Chaps. 1 and 2.
[2] Huxley, J., *Religion without Revelation* (Ernest Benn Ltd, London, 1927).
[3] Barr, J., *Biblical Faith and Natural Theology* (Clarendon Press, Oxford, 1993), esp. Chap. 6.
[4] *Hosea,* Chap. 3.
[5] *Genesis*, Chap. 22.
[6] Dawkins, R., *The God Delusion* (Houghton Mifflin, Boston and New York, 2006), p. 242.
[7] Kierkegaard, S., 1843, *Fear and Trembling*, English translation by W. Lowrie (Princeton University Press, 1941).

8. *I Kings*, Chap. 12, v. 5; *II Kings*, Chap. 16, v. 3, Chap. 21, v. 6.
9. *Isaiah*, Chap. 6
10. *Revelation*, Chap. 1.
11. Poincaré, H, *Science et Méthode*, Paris, 1914; authorized English translation of this and two other works under the title *The Foundations of Science* by B.G. Halsted (The Science Press, New York, 1929), pp. 387-388.
12. Polyani, M., *Personal Knowledge* (University of Chicago Press, corrected edn., 1962), pp. 305, 314.
13. Freud, S. *The Future of an Illusion*, 1927; English translation by W.D. Robson-Scott (The Hogarth Press, London, 1928), Chap. 3.
14. Wilson, D.S., *Darwin's Cathedral* (University of Chicago Press, 2002).
15. Wilson, D.S., and Wilson, E.O., *Evolution "for the Good of the Group"*, American Scientist, **96,** 380-389, 2008.
16. Stenmark, M., *How to Relate Science and Religion: A Multidimensional Model* (Wm B. Eerdmans Publishing Co., Grand Rapids, Michigan, 2004), p. xi.
17. Hayden, B., *Shamans, Sorcerers and Saints* (Smithsonian Books, Washington, 2003), p. 12.
18. *Psalms*, 14, v. 1 and 53, v. 1.
19. *Psalms*, 73, v. 11.
20. *Rig Veda*, 10: 129, v. 7. (I am indebted to Harold Coward for this reference.)
21. Brooks, M., *New Scientist*, 7th February, 2009, pp. 31-33.

CHAPTER 8
Imagery, Miracles and Prayer

> Speak to Him thou for He hears. And
> Spirit with Spirit can meet –
> Closer is he than breathing, and
> Nearer than hands and feet.
>
> Alfred, Lord Tennyson, *The Higher Pantheism*

In earlier chapters I have argued that belief in God is not necessarily irrational, as many scientific materialists seem to believe. In this Chapter I shall look at three elements of religion that do, at first sight, appear to be irrational: religious imagery, the question of miracles, and the practice of prayer.

The imagery that has grown up around conventional religion is often a stumbling block for our modern minds. That imagery is ancient and pre-scientific and much of it no longer carries conviction. There may have been a time when talk of pearly gates and the idea of playing harps somewhere "up above the bright blue sky" for all eternity was both credible and inspiring. Likewise, the thought of being consumed for all eternity in the sulphurous flames believed to be inside the Earth was credible enough to inspire fear. Now, all that imagery is often laughed at, and is found mostly in cartoons that are not very funny. Leaving aside, for the moment, weightier questions about the ideas of judgement and reward or punishment in an afterlife, which will be discussed in Chapter 10, the imagery in which those ideas have traditionally been expressed seems ludicrous today, except for those people who believe that because some sanction for the imagery is to be found in Scripture, it must be believed. I have already commented (in Chapter 5) on the imagery of angels. That traditional imagery certainly has a scriptural warrant in the famous vision of Isaiah in the Jerusalem Temple[1], which I discussed in Chapter 7. The account of that vision is a stirring one and I see no reason to doubt that Isaiah "saw" what he said he did, but he was clearly in a trance-like state at the time. His subconscious mind no doubt clothed the reality in an imagery that made sense to him; *we* do not have to believe in humanoid figures equipped with three pairs of wings, as well as the usual complement of arms and legs. There may be, as I suggested in Chapter 5, higher orders of being of which we can be only dimly aware. There is nothing inherently irrational in being at least open to that possibility, but to insist on the imagery is to miss the point.

Imagery and metaphorical language are not the exclusive preserves of religion; plenty of both are to be found in science. Astronomers frequently speak of the birth, life and death of stars, even of "generations" of stars. These biological metaphors are natural and apt, but astronomy is not part of biology and astronomers do not believe that the stars are alive in the sense that we can say even a virus is. Again, in the early days of the study of atomic structure (about a century ago) the atom was often likened to a miniature solar system. The nucleus, in which reposes almost all the mass of the atom, was analogous to the Sun, while the much less massive electrons were seen as analogous to the planets and were conceived of as revolving in definite orbits around the parent nucleus. We have known for a long time that this image is inadequate, even sometimes misleading, and quantum physicists prefer to talk about the energy levels of the atom, rather than the orbits of the electrons. Nevertheless, I doubt if I am the only astronomical spectroscopist who often lapses into using the orbital imagery originally created by Niels Bohr, because it does help us to visualize how the lines that we study in the spectra of stars are formed. Scientists tend to call such images "models" – perhaps in the hope that that will sound more respectable!

In fact, the physics of sub-atomic particles provides us with one of the most striking uses of imagery in modern science. Quarks, the particles that make up other particles, such as the proton and neutron, that were formerly believed to be themselves fundamental particles, are spoken of as having properties like "flavour", "colour" and "charm", although we all know that individual particles cannot have the properties that those words usually signify. The terms were applied to quarks, partly because of the whimsical sense of humour of some of the original investigators, but also because they do describe real properties of the particles and some terms had to be found. Either new "learned" words had to be invented, probably derived from Greek roots – unfamiliar and therefore obscure to everyone – or common, everyday words had to be given a new, specialized and metaphorical sense. Particle physicists use imagery for precisely the same reason that theologians do: each group is trying to speak of what is beyond our everyday experience and, therefore, ineffable.

Another particularly striking use of imagery in the physical sciences is the deliberate adoption of one of the oldest known religious symbols, the *ouroborus*, or serpent eating its own tail, which comes to us through Greece from ancient Egypt. A detailed example of the use of this particular image is given by Bernard Carr[2] in the compendium *Universe or Multiverse?*, already cited in Chapter 4. Following the Nobel Laureate Sheldon Glashow, Carr uses this image to illustrate the convergence (or "consilience", to use the term popularized by E.O. Wilson[3]) of research into the fundamental particles of matter and into the very earliest times of the universe (or multiverse). There is an irony here, in that the idea of a multiverse, as explained in Chapter 4, has been developed partly in order to avoid the theistic implications of the apparent fine-tuning for life of our universe, whereas the ouroborus initially signified the constant mutual transformations of the material and spiritual into each other. This is perhaps the only example of an explicitly religious image being used in modern science!

There is also imagery in the biological sciences, especially in some of the popularizations of them. The most striking example is the "Selfish Gene"; an individual gene can no more act selfishly than can an individual quark exhibit charm, although there is perhaps more justification for that particular metaphor in biology than for the one in physics. "Selfish Gene" is a striking image and appears to have resonated with a wide reading public, even if not all evolutionary biologists are happy with it. In fact, Richard Dawkins owes much of his deserved success as a popularizer of science to his ability to come up with such vivid images – *The Blind Watchmaker*, *River out of Eden* and *Climbing Mount Improbable* are others that he has used as book titles, and his concept of the "meme", a thought that is capable of spreading from one mind to another in a way reminiscent of the spread of an infectious virus, is also an image. These images are fresh and help people to understand concepts that they might otherwise find difficult, but they have their limitations and there is always the danger that they will eventually grow stale and seem as trite and unconvincing as much traditional religious imagery seems to so many people today. Indeed, since I first wrote this paragraph, Elsdon-Baker has suggested that the time has already come to stop talking of the "Selfish Gene"[4].

Of course, there are also differences between the scientific and the religious uses of imagery. Scientists will change their imagery once the older images are seen to be not only inaccurate but misleading (as we saw above, with Bohr's image of the solar-system atom). I doubt very much that particle physicists, or their successors, will be speaking of colours or flavours of quarks a century from now. The younger ones, in that future time, may even be puzzled that the scientists of our era could talk in that way at all seriously. I doubt, too, if the cosmologists of the future will refer to the "Big Bang" as often as we do. Again, scientists are well aware that they are using imagery, as, I believe, are most of those who read the popular accounts of modern scientific theories. All are aware that the image and the reality are different. Religious imagery, on the other hand, tends to become hallowed by tradition, and while professional theologians are perfectly aware that they are dealing with imagery, many people confuse the image with what it is meant to symbolize. If religious believers of a certain kind have an experience which, like Isaiah, they interpret as a vision of angels, they are likely to see the angels as Isaiah saw them. Convinced of their own experience, such people are likely to resent any expression of scepticism about the imagery, perceiving it either as disbelief in the reality behind the imagery, or even as an attack on their own veracity.

As I observed in Chapter 3, the Second Commandment is that we should not make any graven image or bow down and worship one, and I pointed out that Christians have not observed this Commandment as punctiliously as Jews and Moslems have. Perhaps this is one passage in Scripture that Christians would have been wiser to take more literally (although allowing for some latitude in the interpretation of the word "graven"). There is always the danger of confusing the image with what it stands for, and that is idolatry. Even a representation as sublime as Michelangelo's fresco, on the Sistine ceiling, of the creation of Adam, can generate confusion and reinforce primitive notions of God as an old man in the sky. Think of the number of cartoons you have seen based on that one fresco, many of which make rather tasteless fun of it. Making "graven images" misleads not only the believers, but also the unbelievers, who may be forgiven for confusing the image and the reality if they have the impression that believers themselves have made no distinction between the two. Thus, although imagery has its place and may sometimes be necessary, it has its dangers, too – as puritanical and iconoclastic movements within religious traditions have perceived.

We move from imagery to miracles. There may not, at first sight, be an obvious connection between the two, but many alleged modern miracles are associated with statues that "weep" or "bleed", or with images that mysteriously appear on walls or windows. I do not wish to discuss such alleged miracles in detail, however, since in most, if not all, such cases, I find myself on the side of the sceptics. From a slightly different point of view, even some of the biblical miracle accounts, for example the turning of water into wine or the calming of the storm, may originally have been intended to suggest mental images that make a point about the person to whom the miracle is ascribed, rather than to be taken as historical narratives. *We* might think it wrong to mix genres in this way, in one book, without warning the reader, but the attitudes of earlier ages were different. The borderline between things that are intended to be read literally, and those that are intended metaphorically is, therefore, sometimes rather hard to discern. Other biblical miracles, such as the feeding of the five thousand, may be accounts of historical events embroidered with pious exaggeration.

Nevertheless, all religions do have a miraculous element in them and since, to many modern scientific minds this is, again, a stumbling block to belief, we do have to confront the problem. In earlier ages, people may well have been persuaded to believe by stories of the miraculous, but to us a miracle is the very denial of rationality. The modern discussion of miracles is firmly rooted in David Hume's[5] discussion of them, and in his argument that the testimony to an improbable event must be so strong that it is harder to believe the testimony false than to believe that the improbable event did actually happen. This condition sets the bar very high; obviously the biblical miracles, for the most part, do not clear it, unless, as Hume conceded (presumably ironically) you believe the Bible to be infallible and literally true. Hume's argument has been reprinted together with several modern commentaries on it in a collection edited by Richard Swinburne[6].

We are back again with the notion of prior probabilities, as defined by Bayes, a somewhat older contemporary of Hume, who, David Owen argues in the collection just referred to, was probably familiar with Bayes' theorem. What makes us assess an event as improbable? Primarily, of course, we call an event improbable if it fails to conform to our previous experience. We do not usually observe water being turned into wine by the simple expedient of pouring it into large earthenware jars and then drawing it off again, or paralyzed patients getting up from their beds and walking, at a simple command, or, most of all, people returning from the dead. If you believe in the inerrancy of Scripture, that trumps experience, but Hume, like us, lived after Newton and that gave him a further argument: experience is determined by the laws of nature. Whether or not Hume ever knew of the hymn quoted in Chapter 4, he would have approved of the line:

Laws which never shall be broken

Even if he might have had difficulty adding:

For their guidance hath He made.

The biblical literalist, of course, can still argue that an omnipotent God can suspend the laws of nature and that the inerrant testimony of Scripture is sufficient proof that this has, on occasion, been done. Of course, we now recognize that many laws are statistical in nature: the highly improbable is not necessarily the impossible. Many things, however, are so highly improbable that we are even less likely to encounter them than we are to be dealt a complete suit in a game of bridge; for all practical purposes, we can consider such things to be impossible and, if they happened, we would still regard them as miraculous. Neither statistical mechanics nor Heisenberg's principle of uncertainty give us much reason to expect the miraculous as a result of the operation of laws of nature. A more important objection to the biblical accounts is the way in which, if they are to be believed, God suspends the laws on some occasions and on others, for which the need seems at least as urgent, leaves events to run their course. A God who turns water into wine at a wedding feast and does not lift a finger to stop the Holocaust seems, to say the least, somewhat capricious.

We should, however, make a clear distinction between the two kinds of biblical miracles: the so-called nature miracles and the miracles of healing. Our generation is perhaps more likely to lend credence to the latter than any generation has been since the Enlightenment. Medically unexplained remissions do occur, even of cancers, and many doctors readily admit that there is much that we do not understand about the workings of the body, about its powers to heal itself, and about the relationship between the mind and the body, with which we grappled in Chapter 6. Earlier, in Chapter 2, I referred to the healings that have occurred at the shrine of Lourdes, nearly seventy of which have been claimed by the Roman Catholic Church to have been miraculous. Records have been kept there of those who claim to have experienced a miraculous healing and their claims have been investigated by medical experts, not all of whom are Roman Catholics, or even religious believers of any kind. According to a report in 2004[7], between 6,000 and 7,000 people over the approximately 120 years that the shrine had then been a place of pilgrimage believed themselves to have been miraculously healed during, or as a result of, a visit to Lourdes, but Church authorities set definite criteria for accepting a cure as miraculous and only 66 cases (about one per cent of the claims) had been accepted as such at that time (by the time of writing, another had been accepted). Whatever criticisms may be levelled concerning the operation of Lourdes, the Roman Catholic Church has made a genuine attempt to compile a scientific record of the cases it regards as miraculous.

Unfortunately, that attempt has not satisfied all those who have inquired more closely into the records. In 1957, D.J. West[8] published a little book entitled *Eleven Lourdes Miracles* in which he discussed the records of eleven cases from 1947 until the date of his writing. He did not go back before 1947, because the system of recording and attesting unusual cures was less formal, and probably less thorough, earlier. He was a physician himself and a member of the Society of Psychical Research and can, therefore, be assumed both to have been qualified to assess the medical records and at least open to the possibility of miraculous healings. Nevertheless, he did not find any of the eleven dossiers he studied to have been fully satisfactory. Either some records were missing or diagnostic tests that should have been made had not been, or alternative diagnoses were not properly considered. The first two of these deficiencies can perhaps be partly explained by the fact that several of the cases studied involved French patients living in France throughout the Second World War. Nevertheless, West is undoubtedly correct in pointing out that they render the claim of "miracle" suspect.

Over half a century has elapsed since West wrote that book and, as we all know because we are all beneficiaries of the fact, medical science has made great progress in that time. More diagnostic tests are available, and probably more widely used. We can at least hope that the Committee that assesses claims has learned from experience and from the criticisms of people like West. It would be useful if some independent and sympathetic medical expert would undertake a discussion of more recent dossiers in the same spirit with which West worked – he simply assessed the dossiers as he would if they were used to support the claims for a new drug and tried not to consider the claimed miraculous element until his assessment was complete. Ideally, one should have diagnostic tests immediately before and after the alleged cure but since it is of the very nature of miraculous events that they are unexpected, this requirement is rarely, if ever, met.

The sixty-sixth claim is instructive, whatever one's final judgment on its nature may be. The subject was a young Frenchman who in 1972, at the age of thirty-six, first noticed neurological problems. In 1984 he was diagnosed with multiple sclerosis and, a year later, the disease had progressed so far that he was confined to a wheelchair. By 1987 he was receiving a full disability pension. Friends persuaded him to go to Lourdes, and while there he noticed feeling returning to his limbs, and a few days later he was completely recovered. The medical superintendent did not even present his case to the examining committee until 1992, and the committee then promptly called for further investigation. Only in 2003 did the committee agree that the cure was genuine and inexplicable in terms of current medical knowledge. The final step, officially declaring a miracle, rests with the bishop of the patient's home diocese: in this case, he acted on the committee's decision. This summary is typical of published accounts of Lourdes' miracles and is at first sight quite impressive. I admit to being impressed by it myself until I read West's book and realized how much more a medically qualified person would want to know.

I can think of at least three lines of argument that the sceptic will bring against regarding this particular event as a miracle, in the popularly accepted sense of the term. The first is that multiple sclerosis is well known to be a disease that frequently allows the patient periods of remission. (West makes this point himself in discussing another case of the same disease, even going so far as to say that complete and permanent remissions are not unknown.) In this case, the remission appears to have been complete and to have lasted well over a decade. Members of the examining committee, however, are selected for their medical expertise, not for any particular religious conviction, so all of them must have been fully aware of the remissions characteristic of the disease and yet were convinced that the recovery of this patient was medically inexplicable.

The second line of argument, of course, is to cite the placebo effect. The same answer given for the first line of argument applies here also; but there is a more important answer. The very existence of the placebo effect surely indicates that the mind can influence the body. As I observed in Chapter 6, if the mind believes that the body is being treated, even though the mind has been deceived, the body sometimes gets better. This is hardly what one would expect to happen if the mind were merely an aspect of a sufficiently complex brain. I do not think that Crick's *Astonishing Hypothesis* would predict that sort of thing, although Crick himself, of course, must have been well aware of it.

The third line of argument is actually the sceptic's best: to attribute the cure to direct Divine intervention in the normal healing process is once again to invoke the "God of the gaps". A cure that is medically inexplicable at the beginning of the twenty-first century may well be explicable by the end. Perhaps a number of presently unknown factors came together in this case, as in other similar cases, and the cure followed as a perfectly natural result. Religious believers should not be too ready to dismiss this argument, but they are entitled to point to the fact that these cures, which even if natural are certainly at present unusual, do seem to occur more frequently at sacred shrines than elsewhere. Approximately once every two years at Lourdes, a cure sufficiently unusual to pass a rigorous medical inspection is added to the list of claimed miracles. I do not doubt that other shrines, in other religions, could point to similar results. Bathing in the Ganges, for example, however unhygienic the practice may appear to our Western minds, may well cure many of those who do it. The unique aspects of Lourdes, so far as I know, are the attempts to keep meticulous records and to examine the claims rigorously. Perhaps the associations of the place work in such a way on some patients that as yet unknown healing powers of the body are able to work. That is a natural explanation that even sceptics should be prepared to consider, and one to which West, who accepted that several of the cures were at least unusual, was feeling his way. Such an explanation, just like the placebo effect itself, is also evidence of a two-way interaction between mind and body – a possibility that should surely be investigated rather than dismissed solely on account of its failure to fit into a materialist world-view. An investigation into why these unusual cures do occur might, at the very least, lead to a medical breakthrough.

Believers, on the other hand, must still be troubled by the apparent capriciousness of the healings. The Medical Director at Lourdes himself estimates that ninety per cent of those who go there with illnesses or handicaps leave in the same condition – although they may feel better because of the fraternal atmosphere and the care that they have received. The conventional religious answer is to say that it all depends on the patient's faith; even Bertrand Russell[9], in an attempt to downplay the significance of healings at Lourdes, suggested that any doctor in whom the patient had faith could probably have effected a cure, thus conceding the power of mind over matter. There is no evidence, however, that the subject of the sixty-sixth miracle had any great expectations of his visit, while, no doubt, many who went to Lourdes with very strong faith came away disappointed. "Faith" alone, therefore, seems to be an oversimplification. After all, St Paul himself[10], whose faith was surely beyond question, had his "thorn in the flesh" that would not go away, despite his earnest entreaties. So, again, the element of capriciousness should be worrying to believers; if the "miracles" of Lourdes are Divine acts, we cannot help wondering why God heals some and turns away so many others.

Thoughts turn naturally from miracles to prayer, which is commonly understood to be asking God (or the gods) for something that one would probably not receive in the normal course of events, whether that be a material good or the resolution of a problem. In so far as this is true (a point to which we shall return) the petitioners are seeking their own personal miracles. Many who are prepared to believe in a deist god, the Creator of the world, or at least an Aristotelian first cause, balk at the notion that God can be swayed by human requests. John Tyndall[11], a prominent Irish physicist of the nineteenth century who was opposed at least to conventional religious beliefs, argued that praying for rain, or for a sunny day, was praying for a suspension of the law of conservation of energy. Although he was careful to avoid saying that such a suspension would be impossible, since, by hypothesis, an omnipotent God could obviously bring it about, he himself clearly did not believe that such a prayer could be efficacious. (He lived before Einstein formulated the famous equation $E = mc^2$.) From our greater knowledge of how our weather is the result of the complex interaction of land, sea and air, we might suppose that praying for specific weather is asking only for a redistribution of energy – something that could surely be achieved even without invoking omnipotence! The problem with that sort of prayer is (as Tyndall also hinted) a moral one, rather than a scientific one. The world is one and our weather cannot be modified without compensatory modifications being made elsewhere. The butterfly in the Amazon basin, whose flapping wings are said to produce a hurricane in the Caribbean, may be a picturesque exaggeration (imagery again!), but our good weather may well be bought at the expense of someone else's bad weather. To pray for a sunny day in order that we might enjoy a picnic is simply selfish. Even to pray for fine weather so that local farmers may harvest good crops is questionable. The only way to avoid moral censure is to pray for suitable weather for the whole world, and to recognize that that sometimes means that we will not have the weather we want, where and when we want it. Even then, there is little use praying if we are not prepared to take whatever steps are in our power to preserve a liveable climate on the surface of the Earth.

That does not, however, answer the question of whether prayer can have any value at all. Atheists, of course, dismiss that notion entirely, and deists find it incredible, but both might be willing to concede the indirect benefit hinted at in the last paragraph: what people care enough to pray about they may also care enough to do something about, and the act of praying focuses the mind for action. Nevertheless, even in the days when the entire universe was believed to be somewhat smaller than we now know the solar system to be, it was bold to suggest that the Creator of the universe could be concerned with each individual human being, as was recognized by the Psalmist who asked "What is Man that Thou art mindful of him?" Given what we now know about the age and extent of the universe, such a suggestion seems outrageously audacious. It is not surprising that many people are unable to believe it, but we refer again to the "Argument from Personal Incredulity": that some people find some propositions hard to believe is not, in itself, evidence against those propositions. There are two sorts of difficulty with the concept of praying for something. One is the kind of difficulty raised by Tyndall in connection with prayers for favourable weather, namely: an answer might require interference with the laws of nature and would therefore be miraculous. The other is whether or not the proposition that prayers will be answered is really consistent with properties ascribed to God: omniscience, omnipotence and perfect goodness. A God with these properties obviously knows what we want and, much of the time at least, wishes to give it to us, unless that same God can foresee that some of the things we ask for would, in fact, be bad for us. Prayer, therefore, seems superfluous.

Two things need to be said because sceptics often appear to overlook them, or even to be unaware of them. First, prayer is not *only* a matter of asking for things (or for help); second, theologians are, of course, well aware of the difficulties mentioned in the last paragraph, and have grappled with them over the centuries. All the classical writers on Christian doctrine have discussed the matter, including Augustine and Aquinas. It did not require post-Enlightenment scepticism to make people aware of these difficulties. Two modern discussions (by Terence Penelhum[12] and Eleonore Stumpf[13]) have been reprinted in the collection by Richard Swinburne that I have already cited. Readers can find their way into the classical literature of the problem from the references given by these two authors. An important part of prayer that is *not* petitionary is thanksgiving. Even the sceptic who does not believe that there is any one to thank must surely recognize that those of us who live in the developed world are extremely fortunate and privileged. Ancient folk wisdom advises us to "count our blessings" and that is a discipline we can all observe from time to time, whatever we do or do not believe about the source of those blessings.

In early religions, however, prayer *was*, very probably, primarily "petitionary prayer"; that is, attempts to persuade God, or the gods, to change their minds or to intervene in the world. The Christian concept of God is an uncomfortable amalgam of the Hebrew concept of a God Who intervenes by "mighty acts" and the Aristotelian concept of a prime mover or first cause, aloof from the world and neither influencing nor being influenced by the concerns of human beings. Nevertheless, petitionary prayer does have a place in Christian prayer. While many are convinced that their prayers have been "answered" there is no way that that conviction can be conveyed to determined sceptics who can always dismiss apparent answers as coincidence and point to the large number of apparently unanswered prayers to back up their case. I do not suppose that there is any example of a prayer that the petitioner believes has been answered that cannot be explained away somehow, but the sceptics who dismiss all such claimed events as "coincidences" are simply the mirror images of the creationists who argue that if human beings evolved from some ape-like ancestor then there ought to be transitional forms, and then dismiss all the transitional forms pointed out to them as "diseased individuals". Perhaps it is not possible to prove such an assertion wrong in any individual case but, just as the argument for evolution depends on the total mass of evidence, so does the argument for answered prayers or miraculous healings. The sceptic ought to take into account the number of "coincidences" associated with a particular activity, while the believer should be troubled by the apparent capriciousness of the Deity Who appears to pick and choose which prayers shall be "answered".

In these circumstances, it is not surprising that some people try to test the efficacy of prayer by statistical experiments, some of which are going on now. One group of patients in a hospital, for example, is prayed for, while a control group is not; and the rates of recovery are compared. I doubt if such experiments will ever be conclusive and, certainly, those conducted so far have not convinced many sceptics. The situation is rather like that found in parapsychology: effects, if any, are only marginally above the limit for statistical significance and the sceptic can quarrel with either the statistical analysis or the experimental protocols – or both. Many refined statistical techniques are now available and even expert statisticians are not always agreed on which technique is the most suitable for a given situation. As for the experimental protocols, who can *guarantee* that *no-one* is praying for any individual member of the control group? Richard Dawkins[14], in *The God Delusion*, enjoys some fun at the expense of a recent experiment along these lines that showed no clear-cut difference between the patients prayed for and those not. He rightly calls into question the whole idea of testing prayer in this way – he could have quoted Jesus's answer to the Devil: "You shall not put the Lord your God to the test", but did not – and claims, I suspect with justification, that if the results had been positive, they would by now have been proclaimed from very many pulpits. Perhaps the wisest words on this sort of approach were penned over half a century ago by Dorothy Sayers[15], in a footnote to her Introduction to her own translation of Dante's *Purgatorio*. She was commenting on the fate of Guido da Montefeltro, whom Dante consigned to Hell because he had relied on an absolution from a Pope given in advance of the actual sin:

> The reason is obvious: grace abounds only when there is genuine repentance, and we cannot, as the logical demon rightly observes (*Inf.* Xxvii, 118-20) simultaneously will sin and repentance, since this involves a contradiction in terms.
>
> Incidentally, a similar logical fallacy attends all ingenious proposals to "test the efficacy of prayer" by (for example) praying for patients in Ward A of a hospital and leaving Ward B unprayed for, in order to see which set recovers. Prayer undertaken in that spirit is not prayer at all, and it requires a singular naivety to imagine that Omniscience could be so easily bamboozled.

I think that Dorothy Sayers, if she were still with us, would be both amazed and dismayed that such singular naivety is still so much in evidence on both sides of the debate. Of course, the recent experiments were double-blind, although it is unclear whether the people praying knew that they were taking part in an experiment. It appears that even double-blind experiments do not bamboozle Omniscience. Incidentally, in the interests of justice we should note that Dante also consigned the offending Pope to Hell.

Prayers for the sick probably comprise the majority of petitionary prayers that are offered and, since they are often offered when the medical prognosis is discouraging, they are, in essence, asking for a small and personal miracle. People who have recovered and those who have prayed for them will, of course, be convinced that their prayers were answered; sceptics will probably invoke coincidence and suspend judgement until there is more evidence. Rational argument is unlikely to change the minds of those committed to either view. Once again, each person will assess the evidence in accordance with his or her own estimates of prior probabilities. Much the same arguments apply to this matter of prayer as to the healings at Lourdes, or elsewhere: believers must come to terms with the many similar prayers that appear to have been unanswered, and sceptics must offer some explanation of why unusual cures are often associated with a specific activity – to wit, prayer.

Another common kind of petitionary prayer is for guidance in a difficult situation in which petitioners find themselves and which renders them uncertain what course of action to take. Prayer can then fulfil the function of calming the petitioner's mind so that rational thought can take over and the best course of action (or at least a good course) becomes obvious. People who pray in such situations may well perceive the successful outcome as an answer to prayer, even though it may have been a consequence of their own rational thinking without the direct involvement of God. Perhaps scientific materialists are so used to concentrated, logical and dispassionate thought that they do not need to resort to the techniques of prayer. As a result, they can easily underestimate the value of prayer for those who would not even claim to be so accustomed, and for whom prayer is not merely asking for help but also a means of quieting their own minds so that they can think more clearly. Once again, this is a situation in which a naturalistic explanation and a theistic one are not necessarily mutually exclusive.

The last paragraph brings us back to the statement at the beginning of this discussion: that prayer, whatever its origins may have been, is not only a matter of addressing requests to the Deity. Stilling the mind is not easy for anyone, but it is often the first step to overcoming difficulties and to realizing that power beyond our normal resources is available in a crisis. Recognizing this is not a matter of religious belief; it is a common experience that, when confronted with a major life crisis, many of us keep going in a way that we should not have, in advance, thought possible. I am not thinking only of "Third Man" interventions that I have mentioned already in Chapters 3 and 6 – these may be the most dramatic manifestations of powers that we do not usually tap – but also of the many lesser occasions when people keep going when confronted by difficulties that do not necessarily involve physical danger. Perhaps these resources are simply within our own being and we fail to use them much of the time, or perhaps they are outside us, waiting for us to "plug into" them. Whatever the truth may be, the great religions of the world have developed techniques for stilling the mind, and they appear to provide their adherents with means to tap the resources needed to meet "the changes and chances of this fleeting world". Theists are inclined to see this as drawing on the strength of God. Buddhists would probably regard it as realizing something of our true nature. A Freudian might simply regard it as allowing the subconscious mind to take over. Perhaps it does not matter all that much which explanation one favours: as remarked above, they are not necessarily mutually exclusive and the resources appear to be available regardless of the imagery we use to describe them. What is surely unscientific, and even foolish, is to dismiss the process as useless without first putting it to the test.

References:

[1] *Isaiah* Chap. 6.
[2] Carr, B., *Introduction and Overview*, in *Universe or Multiverse?* (Cambridge University Press, 2007), pp. 1-29.
[3] Wilson, E.O., *Consilience: the Unity of Knowledge* (Alfred A. Knopf, New York, 1998); reprinted in Vintage Books (Random House, New York, 1999).
[4] Elsdon-Baker, F., in *New Scientist*, 18th July 2009, pp. 24-25.
[5] Hume, D., *Of Miracles*, in *An Enquiry Concerning Human Understanding*, 1748 (reprinted on pp. 23-40 of reference 6).
[6] Swinburne, R., (ed.), *Miracles* (Macmillan, New York, 1989); see especially the article by D. Owen, pp. 115-132.
[7] *Manchester Guardian Weekly*, Oct. 8-14, 2004, p. 21.
[8] West, D.J., *Eleven Lourdes Miracles* (Helix Press, Great Britain, 1957), pp. x + 134.
[9] Russell, B., *What is an Agnostic?* Originally published 1953 in *Look* Magazine, reprinted in *The Basic Writings of Bertrand Russell*, R.E. Egner and L.E. Denonn (eds.), (Simon and Schuster, New York, 1961) pp. 577-584.
[10] St Paul, *Second Epistle to the Corinthians*, Chap. 12, v. 7.
[11] Tyndall, J., *Fragments of Science* (P.F. Collier and Son, New York, 1902), Vol. 2, pp. 260-286.
[12] Penelhum, T., see ref. 6, pp. 153-166.
[13] Stumpf, E., see ref. 6, pp. 167-188.
[14] Dawkins, R., *The God Delusion* (Houghton Mifflin, Boston and New York, 2006), pp. 242-3.
[15] Dorothy L. Sayers in *Introduction* to *Dante The Divine Comedy: II, Purgatory* (Penguin Books, Harmondsworth, 1955), footnote on pp. 68-9.

CHAPTER 9
The Historical Relations of Science and Religion

> Grau, teurer Freund, ist alle Theorie
> Und grün des Lebens goldner Baum.
>
> Johann Wolfgang von Goethe, *Faust, Part 1*

Considering that Goethe is reported to have said that he prized the discovery of his theory of colour above the creation of Faust, the mixture of colours in the above quotation is rather strange. We owe the anecdote to Eckermann, but I sometimes wonder if the great man was not above pulling his faithful interlocutor's leg! However that may be, Goethe was not expressing his own view but trying to put words into the Devil's mouth. The "theories" that concern us here are two, each held by many people, about the relations between science and religion, theories that are not so much concerned with shades of grey, let alone hues of green and gold, but with the stark contrast of black and white. The first theory holds science to be pure reason and knowledge (white) and religion to be "blind faith" and obscurantism (black); the second, a mirror-image of the first, holds that science is godless, perhaps even satanic, hubris (black) and religion true humility (white). Those who adhere to either one of these theories have one thing in common: the belief that conflict between science and religion is inevitable and each of us must choose one or the other.

Mention the issue of science and religion in conversation, and before very long one or both of two names will crop up: Galileo and Darwin. Indeed, we have had several occasions to refer to them already in this book. The series of events that have made these two men particularly famous in this context are themselves often seen in black-and-white terms. As we saw in Chapter 1, modern historical research has shown that each of the "real-life" stories centred on these two men were much more complex than is generally believed: green and gold are more appropriate colours to describe the stories than are black and white. Those stories were also atypical. Heilbron[1], in his book *The Sun in the Church*, makes clear how, especially in Italy, cathedrals and other large churches were often used as astronomical observatories. A suitably placed hole high up in the wall would admit a beam of light, permitting measurements of the meridian altitude of the Sun throughout the year. The Church encouraged this work in order to improve the ecclesiastical calendar; the astronomers used the buildings to investigate the apparent motion of the Sun more closely. There was thus a mutually profitable symbiosis between science and religion, and, in fact, the Roman Catholic Church in Italy was one of the biggest patrons of astronomical research in that country *at the very time* that it was proceeding against Galileo.

Early modern scientific research (as we call it; its practitioners called it philosophy) was often religiously motivated. Copernicus was a cathedral canon; Newton, Kepler and Galileo were all religious men, even if they were each, in different ways, somewhat unorthodox. I strongly suspect that part of the reason for Galileo's recantation was that he really did believe that the Church held the keys of Heaven and Hell. He may well have feared the threat of torture (although the Roman Inquisition was not allowed to carry out that threat on a man of Galileo's age) but he feared still more dying excommunicate, even though he could sometimes be lax about religious observances.

More important than the behaviour of individuals when they fell foul of ecclesiastical officials, however, was the motivation of those same people for undertaking their research. They believed that God had created the world and that, in trying to find out how that world worked, they were glorifying God, every bit as much as if they had chosen a life of prayer and contemplation. In a metaphor much used in those early days of what we now call modern science, there were two books: Scripture and Nature. Both books were written, or at least inspired by God, Whose character, it was believed, was revealed in different ways in each: we can learn from either. Although human authors can easily contradict themselves, even in the same book, the Divine Author presumably remains consistent in both books. Apparent contradictions, therefore, arise only from human misunderstanding of one or both of the books. Galileo's greatest mistake was to undertake, in the matter of Joshua's miracle, to show how the then accepted literal reading of Scripture was such a misunderstanding. By doing so, he crossed the line between the area of knowledge in which he could claim to be expert, and that in which the clergy were the acknowledged experts, and thus became vulnerable to their censure.

Perhaps one of the most obvious examples of attempts to reconcile the two books was Thomas Burnet's[2] (1635-1715) *Sacred Theory of the Earth*. To most modern minds, the book reads like a fantasy: Noah's flood is taken quite seriously as a major epoch in the Earth's history and Burnet describes sudden shifts in the Earth's axis of rotation to produce climate changes that would quite dwarf those that are the focus of our present-day concerns. (Burnet's older contemporary, John Milton[3], expressed a similar idea in *Paradise Lost*.) We may dismiss this as nonsense, but in terms of the science of his day, such changes were not considered impossible. Even Newton, whose opinion Burnet sought, wrote quite favourably about the idea. The important thing was that Burnet at least envisaged that the Earth *had* a history: to that extent, he can be seen as a forerunner of the science of geology, although another of his contemporaries, Robert Hooke, came much closer to modern ideas, as Allan Chapman[4] has shown in his recent biography of the latter.

The symbiosis between science and religion continued into the eighteenth century, but, during this period, the growing importance of Enlightenment thought and the related rise of deism began to wear it down, particularly through a growing questioning of the authority of the Bible and of the accuracy of its purported history and science. That questioning did not arise solely from the new scientific ideas. The Enlightenment also affected the way in which the Bible was viewed, even by theologians and biblical scholars. In particular, within the German Lutheran Church, modern biblical criticism began to emerge, leading to the so-called "higher criticism" of the nineteenth century. The insistence of the first fundamentalists on biblical literalism, in the early twentieth century, was at least as much a reaction against the interpretations of biblical critics as against the theories of the scientists. I am told by those with more teaching experience than I have that it is still true that students inclined to interpret the Bible literally are more disturbed by biblical criticism than by modern science, and it is at least possible that the present strength of fundamentalists is partly a reaction to the very radical biblical criticism of groups like the Jesus Seminar[5].

The age of the Earth first became an issue between the scientists and the religious believers of Europe during the Enlightenment. Although, as we saw in Chapter 1, Hutton and Lyell were primarily responsible for enlarging our notions of how long the Earth had existed; the discovery of fossils by the Dane, Steno, in the late sixteenth century led his English contemporary, John Ray whose life and work have been described by C.E. Raven[6], to speculate on their implications for estimates of the age of the Earth, if they were indeed fossilized remains of extinct creatures. Before that time, there was no conception, in Western thought, of the Earth having had a history separate from that of the human race. The Earth was supposed to have been created pretty much in its present form, and populated from the beginning with the species that we now know. Of course, people were aware of earthquakes, volcanoes and floods, but, apart from The Flood, these were seen as local affairs that had relatively little effect on the Earth as a whole. The Bible and the works of Hesiod and Homer were the only documents in the West that even purported to tell us about the earliest times, and since the first of these was "Holy Writ" there was no reason to doubt the essential truth of the account of, or even the approximate date for, the Creation that it seemed to imply. Although, as again we saw in Chapter 1, there were several different attempts to deduce the date of the Creation of the Earth, they all agreed in making the Earth much less than 10,000 years old.

The possibility that fossils were indeed the remains of animals, some of which belonged to species that are now extinct, began to chip away at that early eighteenth-century consensus that the Bible was a reliable guide to ancient history and natural history. Neither of the biblical accounts of creation refers explicitly to the possibility of extinction and, indeed, they are generally seen as at least implying that all species were created together at the beginning, having neither been added to nor subtracted from since. If, however, some species had died out, why should others not have been created since to take their place? Could existing species have been modified from their earlier forms? Thus, the discovery of fossils and the understanding of their significance opened the way to the formulation of the concept of evolution, or transmutation of species, as it was often called in those earlier days. Diderot and Erasmus Darwin (grandfather of Charles) were among those who entertained early theories of what we would now call evolution. Diderot's philosophical development took him from religious belief, through deism, to atheism, and Erasmus Darwin was probably a deist, too: so, early on, evolution became associated with at least some degree of religious scepticism, a fact that may have helped to create the tension between belief in evolution and religious orthodoxy that many people still feel.

Perhaps the most important evolutionist before Charles Darwin was, however, the French naturalist Lamarck, who developed a coherent theory of evolution which Darwin himself was prepared to incorporate within his own theory. Like Darwin, Lamarck believed that plants and animals changed by adapting to their (possibly) changing environment. He supposed that the organism could change according to its *needs*. Lamarck himself used the French word *besoins*, but Cuvier, a strong opponent of evolutionary ideas, who wrote the official *éloge* of Lamarck for the *Académie des Sciences*, used not only the word *besoins*, but also the word *desirs*. Unfortunately, the first English translation emphasized the second term and spoke of animals *wishing* to achieve changes in their organs – thus besmirching Lamarck's reputation in English-speaking countries even more than it had been in his native France. It is usually supposed that the important feature in Lamarck's version of evolutionary theory was that changes acquired by an individual in the course of its lifetime could be passed on to its offspring – the so-called inheritance of acquired characters. That sort of inheritance was a reasonable guess at the time, although it is emphatically not part of neo-Darwinism, even though Darwin himself considered it to be one of the possible mechanisms of evolution. (It is important to remember that neither Lamarck, nor Darwin, nor any of their contemporaries, knew of the modern theory of heredity, which made both theories vulnerable throughout the nineteenth century.) In my days as a graduate student, however, I heard occasional lectures from Professor Hugh Graham Cannon who maintained strongly that Lamarck had been misrepresented, not only in the matter of wishes and needs, but, more importantly, in the nature of his theory. The lectures I heard were a condensation of his book *Lamarck and Modern Genetics*[7], written in a polemical style that will not appeal to everybody. Some of Cannon's arguments must certainly need revision now that nearly half a century of molecular biology has passed since his book was written, but I suspect that he read Lamarck more thoroughly than most of his contemporaries or successors (in biology) have done, and that his assertions about what Lamarck actually said are reliable. According to Cannon then, the central point of Lamarck's theory was that if there was a need for a particular organ in an animal then that organ would appear. He did indeed go on to say that that organ would be inherited by the animal's progeny, but, Cannon argues, that was a redundant feature in the theory, since if an organ could appear in response to a need in one generation, so it could again in the next.

Since the discovery of the structure of DNA, what Francis Crick called the "central dogma" of molecular biology has played an important role. That "dogma" states that changes in an individual's DNA can affect its proteins, but changes in the proteins cannot affect the DNA. If this is true without exception, then acquired characters cannot be inherited and the central dogma is a restatement in terms of molecular biology of Weismann's idea (which also calls forth polemics from Cannon) of the complete separation of the "germ plasm" from the rest of the body. At the level of popular understanding of evolution, however, I doubt if Lamarckism is dead yet and, until the rise of what Julian Huxley called the "modern synthesis" (of neo-Darwinism and Mendelian genetics) in the 1930s, Lamarck's ideas still enjoyed some support. At that time, as Bowler[8] has pointed out, many Christians felt more at home with Lamarckism, since inheritance of acquired characters (and still more Cannon's interpretation of Lamarckism) seemed less of a chance matter than did random variations acted on by natural selection, and allowed more room for Divine direction. The prominent Anglican scholar and clergyman C.E. Raven was one whose understanding of evolution (which he fully accepted) appears to have been Lamarckian. Ironically, in the Soviet Union, Lamarckian ideas enjoyed a prolonged lease of life, right into the 1960s, in the now completely discredited theories of T.D. Lysenko, who was able to persuade Stalin that inheritance of acquired characters was more in accord with dialectical materialism than was the modern synthesis. Thus, in the mid-twentieth century, the late eighteenth-century ideas of Lamarck were pressed, with complete impartiality, into the service of both theism and atheism!

In standing up for Lamarck, Cannon was certainly not advocating anything like "Intelligent Design". If he had lived to see the concept emerge under that name, he would, I am sure, have been scathing about it – but he might have been just as scathing about "selfish genes"! As one who is not a biologist, I sometimes wonder if the great advances made in genetics have not distracted molecular biologists from the importance of considering the living organism as a whole. Cannon was not alone in emphasizing the importance of that sort of consideration; in that respect he was typical of his generation, the generation of those who had established their scientific credentials just before the discovery of the structure of DNA. Eccles[9], in the book by himself and Popper that I cited in Chapter 6, made a similar point in talking of the growth of the brain in our pre-human and early human ancestors:

I believe that this growth did not arise spontaneously in some kind of uncaused manner, but that it rose in response to the needs, the demanding needs, of the linguistic developments and all of the associated creative aspects in thought, in discursive thought, in critical thought and so on.

A third biologist of that generation to espouse the same point of view was Sir Alister Hardy[10]. Again, he was no Lamarckian in the generally accepted sense of the term, but in his book *The Living Stream* he points to Lamarck's idea that the *habits* of animals would lead to some variations of the kind that we would now call genetic being favoured over others. Animals might seek out an environment more suited to their habits, or even modify to some extent their immediate environment. Thus, Hardy argues, the animal's behaviour can indirectly influence its evolution and augment the action of natural selection.

Distinguished as Cannon, Eccles and Hardy were in their day, we might be tempted to dismiss these ideas of theirs as the ideas of people who had not fully absorbed the implications of the new molecular biology that arose after they had made their marks on science. There are some signs, however, of a return to their point of view, which has been submerged, at least in popularizations of biology, during the last half-century. I referred in Chapter 4 to the recent book by J. Scott Turner[11], *The Tinkerer's Accomplice*. I do not think that Turner would want be called a Lamarckian, any more than would Eccles or Hardy, but, like them, he expresses ideas that are entirely in accord with what Cannon claimed that Lamarck actually said. Towards the end of his book Turner writes:

> ...can evolution in any way be driven by intention? The Central Dogma of Darwinism states that it simply cannot be. I, however, have argued throughout this book that it just might be. At the very least, it seems unscientific to exclude the possibility altogether...

He comes to his conclusion because, as a physiologist, he considers the whole organism and sees natural selection acting at that level, rather than the level of individual genes. There is, in fact, a growing interest in what has become known as *epigenetics*, the study of how an organism's genes and life experience interact in the development of the individual, as described in a recent article in *New Scientist*[12]. Thus, even among biologists who are totally committed to Darwinian evolution as a paradigm within which to work, there is not the unanimity that some popularizers would have you believe that there is no possibility of intention at least sometimes directing the process. It is a mistake to suppose that biologists, except for a small minority with a religious agenda, are united in a strict insistence on natural selection being the sole agency in evolution (with minor roles being allowed for mechanisms such as sexual selection). That insistence may be the majority view, but there is a broad spectrum of opinions, and many who would certainly not accept the designations of "creationist" or "design theorist" nevertheless suspect that processes other than natural selection are involved.

The work of Hutton (1736-1786) and Lyell (1797-1875) on the foundations of modern geology, to which I referred in Chapter 1, was contemporary with, or at latest shortly after, that of Lamarck (1744-1829). Before Hutton published his theory, Abraham Gottlieb Werner (1749-1817) had argued that all rocks were sedimentary in origin and founded what came to be known as the Neptunian school of geology. Volcanoes and earthquakes, Werner believed, perhaps partly in reaction to Anton-Lazzaro Moro who had argued that all rocks were igneous in origin, played only a trivial role in the formation of the Earth's surface (ice ages were not recognized until the early nineteenth-century work of Louis Agassiz). Hutton was amongst those who recognized that volcanoes and earthquakes played a more important role and who became known as Vulcanists. Hutton's great contribution, however, was the introduction of the concept that William Whewell christened *uniformitarianism*, by which he meant that the Earth's surface features had been produced by the same processes that we observe now (sedimentation, atmospheric and aqueous erosion, earthquakes and volcanoes) acting *at the same rate and intensity* that we experience now. In this matter, Lyell followed him. Here was a very direct challenge, both to the biblical chronology and to the notion of a worldwide flood, because the doctrine of uniformitarianism required a much older Earth than a literal interpretation of the Bible allowed. Although neither Hutton nor Lyell (who were probably both deists) explicitly claimed an infinite age for the Earth, they spun their theories as if they could have all the time they wanted. Hutton, in a famous phrase, did refer to the Earth's surface as showing "no vestige of a beginning, no prospect of an end". Not everyone accepted this view, and the rival school of *catastrophists*, particularly associated with Cuvier, held that the Earth's surface developed through a series of catastrophic floods, of which the biblical flood was accounted to have been one.

There was clear potential in the work of Hutton and Lyell for conflict with established religion, but the fiercest critic of those two, as we saw in Chapter 1, was a fellow scientist, William Thomson, later Lord Kelvin, although he may have been partly motivated by religious considerations. We have discussed his arguments in Chapter 1 and seen that, although they were based on the best physics of the day, they were one by one demolished as new discoveries showed ways in which Kelvin's objections could be met. Some of those new discoveries depended on the rise of chemical atomism, stemming from the work of John Dalton, which was published very early in the nineteenth century, and leading to the discovery of radioactivity at the end of the same century. Thus, the scientific aspects of arguments about the age of the Earth were settled by the development of science itself. Ironically, we have now returned to a kind of catastrophism with our belief that the Earth's surface has also been affected by collisions with asteroids and comets, at least one of which collisions may have changed the course of the evolution of life (by wiping out the dinosaurs). In a larger sense, however, these catastrophes are part of the natural order and, statistically at least, occur at a more-or-less constant rate. Such catastrophes, therefore, do not by themselves rule out Hutton's idea of uniformitarianism, although his assumption that all natural processes have maintained a constant rate throughout the Earth's history is no longer tenable.

Darwin published *The Origin of Species* just over halfway through the nineteenth century (1859), right at the height of Kelvin's arguments with the geologists. Darwin knew, therefore, that he would face *scientific* opposition as well as *religious*. The long timescales that the geologists were arguing for were essential if evolution by natural selection was to be possible. Remember that Darwin's reading of Lyell's *Principles of Geology* during the voyage of the *Beagle* was one of the formative influences on the theory of evolution. Thus, in terms of the scientific knowledge of the time, Darwin's theory was vulnerable at two points at least: this question of the available time, and the lack, that we have already noticed, of a good theory of heredity. Because of these problems, scientific support for Darwinism languished after the initial enthusiasm, but revived with the rediscovery of Mendel's laws and the discovery of chromosomes. These discoveries came at about the same time as the discovery of radioactivity; that is, at the end of the nineteenth century, so the force of Kelvin's arguments against the geologists was diminished just as some of the specifically biological arguments against Darwinian evolution were also being removed. By 1941 DNA had been identified as the likely carrier of heredity, when the modern synthesis of genetics and Darwinism was already emerging; a synthesis that was, of course, greatly strengthened by the famous discovery in 1953, by Watson and Crick, of the structure of that molecule.

Looking back at Darwin's own time, however, we see that the scientific situation was then far from being clear-cut. Darwinian evolution was not a triumphal theory being held back by religious obscurantists (any more than Copernican heliocentricism had been); it was a new and exciting theory with potentially very great explanatory power, but vulnerable to at least two scientific objections, which seemed cogent at the time. In some respects, it is surprising that the theory raised any religious hackles at all. Any religious believer who suspended judgement until the scientists had sorted the matter out to their own satisfaction would have been acting quite rationally and would have been above criticism. Of course, some did rush to criticize and we have the famous story of the debate between Bishop Wilberforce and T.H. Huxley (although working out exactly what did happen there is not easy; once again, the real-life story is more complex and multicoloured than the grey, or even black and white, of theory) but many prominent churchmen were just as quick to embrace the theory – the more remarkable since sitting on the fence was, for once, a respectable option. Among those who were open to Darwinian ideas were the future Cardinal Newman, as we know from his private correspondence, a future Archbishop of Canterbury, Frederick Temple, and Darwin's personal friend, Charles Kingsley (see Olaf Pedersen's[13] *The Book of Nature* for a discussion of the attitudes of the first two of these men and Gertrude Himmelfarb[14] for an account of Kingsley's reactions). The religious and scientific communities were each split in their reaction to Darwinian ideas.

Separate from the public reaction to Darwinism was the effect that his theory had on Darwin himself. His was a rather slow progression from conventional, somewhat evangelical believer, quite willing to entertain the idea of becoming a country parson, through doubt and deism to at least an agnosticism like Huxley's, if not complete atheism. Undoubtedly, his scientific knowledge played a role in this progression; he could not see that an omnipotent and loving God would permit the infliction of suffering by one species on another that he saw in nature. But there were other factors that probably would have acted whatever his walk in life had turned out to be. The suffering and death of his favourite daughter was one. Another, as we now know from his autobiography, was the insistence of some clergy that those who doubted would be subject to everlasting punishment. That would mean that Darwin's father, brother and many of his best friends would be so condemned, and that, wrote Darwin[15] in what was (for him) unusually strong language, "is a damnable doctrine." He was at one there with his friend Huxley[16] who considered one of the worst features of all the churches to be their insistence "that honest disbelief in their more or less astonishing creeds is a moral offence, indeed a sin of the deepest dye, deserving and involving the same future retribution as murder and robbery." Even so, Darwin's criticism of the attitude to disbelief was omitted from the first printing of his autobiography, on the insistence of his widow. Her reason was not that she disagreed – her Unitarian upbringing had imbued her with much the same view of the notion of eternal punishment – but that she felt that her husband had failed to realize that the emphasis of Christian teaching had changed during his lifetime, and that his reputation would be harmed if he were seen to be attacking doctrines that many clergy were no longer preaching.

Thus, both the scientific and ecclesiastical worlds were divided in their reception of Darwin's ideas. The term "fundamentalism" had not yet been coined in the late nineteenth century, but divisions were appearing between those who accepted the results of biblical criticism and those who insisted that what *could* be taken literally *should* be so taken. Darwin did, however, discredit at least some of the more extreme forms of the design argument, if not all *biological* forms of it. He removed even the need for such arguments by proposing a dynamic hypothesis for the origin of adaptations (natural selection) in place of the static one of Divine fiat. As far as I can see, this change is irreversible, which perhaps explains the eagerness of some to reintroduce Divine fiat as "Intelligent Design" and, equally, the insistence of Darwinians that that is an unscientific concept. Whether or not the design argument has been as conclusively eliminated from physics and cosmology is, as we saw in Chapter 4, more debatable.

The discovery of the structure of DNA, and the possibilities of genetic modification that that discovery has opened up, together with the increasing creationist reaction to the whole of evolutionary theory, may give the impression that the biological sciences dominated the relations between science and religion in the twentieth century, and did so, once again, largely in terms of conflict, or at least confrontation. That may be true of the latter half of the century, but in the first half physics underwent two tremendous revolutions, the consequences of which we are still working out. Those revolutions were, of course, the development of the theory of relativity and of quantum theory, both of which began in the first decade of the century. According to Eddington[17],

Einstein was once asked by the then Archbishop of Canterbury what religious implications the theory of relativity had and received the hurried reply, "None. Relativity is a purely scientific theory, and has nothing to do with religion." Possibly Einstein wished to distance himself from any confusion between the physical theory of relativity, on the one hand, and philosophical or religious relativism on the other, or perhaps to avoid any confusion that might arise from his treatment of time as a fourth dimension, but, as Eddington pointed out, Darwin could have said the same of the theory of evolution. In denying the absolute nature of space and time and insisting on the role of the observer, Einstein was obviously raising philosophical questions that might well have repercussions for religious believers. The greater upset to established beliefs, however, has undoubtedly come from quantum theory, with its calling into question the very nature of causation and its reintroduction of the notion of action at a distance, which Einstein thought he had got rid of in the macroscopic world of general relativity.

Quantum theory, as much as relativity theory, has forced us to reconsider our ideas about space and time. The phenomenon of quantum entanglement, in which two particles can be so essentially linked that, even if they are at a great distance from each other, they can affect each other instantaneously, shows that our macroscopic ideas of space and time do not apply at atomic and sub-atomic levels. Einstein, of course, was never convinced that quantum theory was a final theory. He believed that the universe must be deterministic – God does not play dice, in one of his famous phrases – and he could not accept the return of action at a distance (or "non-locality" as it is now often called) into physics. Most physicists now believe that he was wrong on both counts, and yet they, too, do not believe that quantum theory is a final theory, since they spend a great deal of time searching for a "theory of everything" – one that will unite our understanding of gravity with those of the other three fundamental forces. Perhaps it is premature, for that reason, to rush into a discussion of the implications of quantum theory for religious belief, but I have already commented on the fact that the quantum-theoretical picture of the atom leads us to a picture of the nature of matter that is much more enigmatic than nineteenth-century ideas were. It is this, more than the theory of relativity, that has led at least some physicists to question the old-fashioned kind of materialism while, somewhat paradoxically, biologists, who after all deal with living organisms, often seem fixed in that same old-fashioned kind of materialism. Our universe does indeed appear to be enigmatic, not only at the level of the smallest particles that make it up, but also at the level of human interactions!

References:

[1] Heilbron, J.L., *The Sun in the Church* (Harvard University Press, 1999); see esp. the Introduction.

[2] Burnet, T., *Telluris theoria sacra*, 1681, translated into English as *Theory of the Earth*, 1684. A modern edition is Thomas Burnet, *The Sacred Theory of the Earth*, with an introduction by Basil Wiley (Centaur Press Ltd, London and Fontwell, 1965).

[3] Milton, J., *Paradise Lost*, 1667, Book X (as arranged in the 2nd Edition), ll. 651-6, 668-671.

[4] Chapman, A., *England's Leonardo: Robert Hooke and the Seventeenth Century Scientific Revolution* (Institute of Physics Publishing, 2005).

[5] Funk, R.W., *Honest to Jesus* (HarperSanFrancisco, New York, 1996).

[6] Raven, C.E., *John Ray* (Cambridge University Press, 1943), Chap. XVI.

[7] Cannon, H.G., *Lamarck and Modern Genetics* (Manchester University Press, 1959), p.51.

[8] Bowler, P.J., *Reconciling Science and Religion: The Debate in Early Twentieth-Century Britain* (University of Chicago Press, 2001), p. 278.

[9] Popper, K.R., and Eccles, J.C., *The Self and its Brain* (Springer Verlag, Berlin, 1978), p. 454.

[10] Hardy, A., *The Living Stream* (Collins, London, 1965), p. 58.

[11] Turner, J.S., *The Tinkerer's Accomplice* (Harvard University Press, 2007), p.224.

[12] Young, E., in *New Scientist*, 12 July, 2008, pp. 29-33.
[13] O. Pedersen, *The Book of Nature* (Vatican Observatory Publications, 1992), (distributed outside Italy by University of Notre Dame Press, Notre Dame, Indiana), pp. 82-87.
[14] Himmelfarb, G., *Darwin and the Darwinian Revolution* (W.W. Norton & Company Inc., New York, 1959), p. 300.
[15] Darwin, C., *The Autobiography of Charles Darwin 1809-1882 with original omissions restored*, edited with Appendix and Notes by his granddaughter Nora Barlow (Collins, London, 1958), p. 87.
[16] Huxley, T.H., *On Agnosticism*, 1889, reprinted in *Thomas Henry Huxley: Selections from the Essays*, Alburey Castell (ed.) (AHM Publishing Corp, Northbrook, Illinois, 1948), pp. 69-91.
[17] Eddington, A.S., *The Philosophy of Physical Science* (Cambridge University Press, 1939), pp. 7-8.

CHAPTER 10

The Harmonization of Science and Religion

> …science without religion is lame, religion without science is blind.
> Albert Einstein, from an address reprinted in *Out of My Later Years* and in *Ideas and Opinions*.

Although many people still seem to think that conflict between science and religion is inevitable and endemic, those familiar with the writings of Ian Barbour[1] will be aware of his four possible modes of interaction between these "two cultures": conflict, independence, dialogue and integration. That scheme has been criticized by Cantor and Kenny[2], who argue that not only the words *science* and *religion*, but also the words Barbour used to relate them (*conflict, independence,* etc.) have changed their meanings in the centuries that have elapsed since the dawn of modern European science, and the scheme is not, therefore, particularly helpful to historians. Another criticism has been made by Stenmark[3], in his recent book to which I have already referred, who particularly objects to the term *dialogue*, pointing out very justly that the other three modes of interaction, even including conflict, inevitably involve some form of dialogue. Nevertheless, Barbour was among the first to insist that the relation between science and religion is not necessarily one of conflict and the first two elements of his scheme, at least, provide a good description of the sequence of events in the English-speaking world during the late nineteenth and early twentieth centuries, in the aftermath of the publication of Darwin's *Origin*.

Although, as I have stressed several times and particularly in the last chapter, the situation was more complex than is generally believed, the work of Darwin and the geologists certainly led to a breach between those, both inside and outside the churches on the one hand, who were willing to accept the new theories, and the more conservative members of the churches on the other hand. There was conflict then, as there had been at the time of Galileo, but that was not the whole story. I have already referred to Frederick Temple's suggestion that the Church would not interfere in purely scientific disputes, if the scientists, on their part, would refrain from pontificating on matters of faith. This, then, was an early statement of what Barbour has called "independence": science and religion deal with different areas of human life in different ways and, therefore, no conflict can arise between them. This idea has recently been revived by the late Stephen Jay Gould[4] in his book, *Rocks of Ages*, under the rather cumbersome title "non-overlapping magisteria", which he abbreviated to NOMA. (The word "independence" seems to me to be much simpler!) There is certainly some truth in this point of view: most scientists do not claim that, *as scientists*, they deal with questions of purpose and value, whatever beliefs they may hold about those matters *as individuals*. There are, of course, a number of well-known and respected scientists who *do* claim that the results of scientific research have shown us that the universe is purposeless and that values are a human creation, and some of those scientists are, at present, quite vociferous. Similarly, many religious believers freely admit that the Bible, or whatever other scripture they prize, was never intended to be a textbook of natural science. Again, there are exceptions, and they, too, are vociferous. Christian fundamentalists do think that where the Bible appears to give information about the natural world, that information is more reliable than at least some of the results and theories of modern science. So the independence doctrine works for those who are willing to accept it, but breaks down before the onslaughts of extremists on either side, who want to dictate how matters should be arranged in the other sphere. As we have seen, Stenmark refers to "scientific expansionists" and "religious expansionists" to describe those people who want to embrace all human knowledge and experience within their own particular ideology or world-view.

A more important objection to the concept of independence was voiced by Sir Arthur Eddington[5], who pointed out (before Barbour had coined his term) that the very idea of separate domains for science and religion contained within it the possibility that there will be frontier areas in which disputes can arise. In the seventeenth century, the relative motion of the Sun and the Earth was one such frontier area. That region ceased to be disputed when the evidence for the Earth's motion (rather than the Sun's) became all but incontrovertible, and even biblical literalists saw that the account of Joshua's miracle could hardly be taken literally. From the time of Hutton and Lyell, the age of the Earth has appeared to some Christians to be a frontier area. Indeed, some are still disputing that frontier area in this twenty-first century, and vociferous groups are lobbying against the teaching of geology and biological evolution in publicly supported schools, at least in the United States, even though many Christians have no problem with accepting either. On the other hand, other areas of modern biological research, stem cells and artificial fertilization for example, are frontier areas for somewhat larger groups of people, because of the ethical issues perceived to be involved. The time may come when these issues will be resolved and it will seem as odd that they could have caused controversy as it does to us that the heliocentric hypothesis could. Meanwhile, it is clear that independence, or NOMA, is not an accurate or full description of the relation between science and religion.

It may have been considerations like this that led Barbour to label his third kind of relationship "dialogue". While Stenmark's contention that *any* relationship, including conflict, involves dialogue is undoubtedly correct, there was a movement in the early twentieth century from the mere agreement that each side should keep to its own sphere of influence to an exploration of the possibility of reconciliation after the disputes of the previous century. The astronomers Eddington and Jeans were prominent in this movement, on the scientific side, as well as the mathematical philosopher Whitehead, and, to a lesser extent, the theoretical physicist Einstein. On the religious side, the names of C.E. Raven, W.R. Inge and E.W. Barnes come to mind; all of them, as it happens, were Anglican clergy of the modernist persuasion, but Barnes, an expert mathematician, had a foot in both camps and was a considerable scholar, although a somewhat controversial bishop.

In my opinion, Eddington was the most profound of those on the scientific side, and he has greatly influenced my own point of view. Nevertheless, professional philosophers criticized him quite strongly, and even Inge was somewhat critical of his views. Eddington was both a Quaker and a mystic and defended a rather individualistic form of religion; as we have already seen, theology did not interest him and, indeed, he was somewhat antipathetic to it. An important study of the relation between his scientific views and religious practice has recently been published by Matthew Stanley[6] (*Practical Mystic: Religion, Science and A.S. Eddington*). Einstein's attitude is more puzzling and hard to assess. He is famous, of course, for his frequent references to God, the best-known probably being "God does not play dice", which I quoted in the last chapter. What is not clear is precisely what he meant by the word "God" (or synonyms such as "the Lord" or "the Old One"). He was a great admirer of Spinoza and much of what he said on the subject of God sounds like pantheism. In an interview[7], however, Einstein was asked if he believed in the God of Spinoza and replied that he could not answer with a simple yes or no. "I am not an atheist," he said, "and I don't think I can call myself a pantheist." Yet he could not believe in a personal God who rewards or punishes His creatures, and he rejected the notion that the individual could survive the death of the body. Einstein's father was a secular Jew, proud of the fact that traditional Jewish rituals had no place in his house. Einstein himself was briefly influenced by religious Judaism, around the age of twelve. After that, however, and for the rest of his life, he consistently denied belief in a personal God, reiterating his denial (according to Max Jammer[8], and confirmed by a recently published letter) even within a year of his death. In 1939, Einstein set out his views on this subject in an address to the theological faculty at Princeton University, and two years later he further expounded those ideas in a contribution to a published symposium (from which the well-known quotation at the head of this chapter is taken). Reading these remarks and others by Einstein, I have wondered if he ever distinguished fully between the notions of a *personal* God and of an *anthropomorphic* God, and if, perhaps, it was the latter that he was primarily concerned to deny.

On just this point, there is an interesting contrast between Eddington and Einstein (the two men knew each other and had a high mutual respect). In his book *Science and the Unseen World*, from which I have already quoted, Eddington[9] tackled the question of a personal God:

> It seems right at this point to say a few words in relation to the question of a Personal God. I suppose every serious thinker is afraid of this term which might seem to imply that he pictures the deity on a throne in the sky after the manner of the medieval painters. There is a tendency to substitute such terms as "omnipotent force" or even a "fourth dimension". If the idea is merely to find a wording which shall be sufficiently vague, it is somewhat unsuitable for the scientist to whom the words "force" and "dimension" convey something entirely precise and defined. On the other hand, my impression of psychology suggests that the word "person" might be considered vague enough as it stands. But leaving aside verbal questions, I believe that the thought that lies behind this reaction is unsound. It is, I think, of the very essence of the unseen world that the conception of personality should dominate it. Force, energy, dimensions belong to the world of symbols; it is out of such conceptions that we have built up the external world of physics. What other conceptions have we? After exhausting physical methods we returned to the inmost recesses of consciousness, to the voice that proclaims

our personality; and from there we entered on a new outlook. We have to build the spiritual world out of symbols taken from our own personality, as we build the physical world out of the symbols of the mathematician. I think therefore we are not wrong in embodying the significance of the spiritual world to ourselves in the feeling of a personal relationship, for our whole approach to it is bound up with those aspects of consciousness in which personality is centred.

In this passage, Eddington makes clear the distinction between a personal God and an anthropomorphic one in a way that Einstein never explicitly did in what he said and wrote, at least in public. Certainly, neither of those two men believed in an anthropomorphic God; quite possibly they both believed in much the same sort of God, but it is very hard to tease out from Einstein's many statements about God just what he did believe. Einstein was sufficiently imbued with his Jewish heritage to abhor idols, and he probably saw anthropomorphic imagery as a form of idolatry, just as the Hebrew prophets did. His denial of belief in a personal God can be seen as a condemnation of idolatry, and yet, paradoxically, so many of his references to God draw on personal terms: only a person can play dice, or be subtle or malicious, or write books. (Einstein once described God by analogy, as the author of a whole library of books!) Perhaps Matthew Arnold's[10] phrase "the eternal something not ourselves, which makes for righteousness" comes as close as anything to what Einstein meant by the word "God".

The most important thing about Einstein in this context, however, is that he never lost his sense of awe *vis-à-vis* the cosmos, and this seems to have been intimately connected with his concept of God. He marvelled that the universe existed at all, that it is rational and ordered, and that we, with our limited minds, can grasp something of that rational order. He wrote once[11]:

> The most beautiful experience we can have is the mysterious. It is the fundamental emotion which stands at the cradle of true art and true science. Whoever does not know it and can no longer wonder, no longer marvel, is as good as dead, and his eyes are dimmed. It was the experience of mystery – even if mixed with fear – that engendered religion. A knowledge of the existence of something we cannot penetrate, our perceptions of the profoundest reason and the most radiant beauty, which only in their most primitive forms are accessible to our minds – it is this knowledge and this emotion that constitute true religiosity; in this sense, and in this alone, I am a deeply religious man.

Almost every scientist could subscribe to these remarks, even those who call themselves sceptics or unbelievers. Both Richard Dawkins[12] and Michael Shermer[13] have criticized the poet John Keats for his accusation that by "unweaving the rainbow" scientists, in particular Newton, took the wonder out of the natural world. Dawkins and Shermer argue that to know the causes of natural phenomena is to increase one's sense of wonder, rather than to diminish it. In this matter, I cordially agree with them. For example, like most people who live within the range of hummingbirds, I enjoy watching the tiny, brightly coloured creatures hovering at the flowers from which they drink nectar. I also marvel, however, that the tiny sips they take can provide more than enough energy to compensate for that spent in the beating of their wings, and I find myself wondering what evolutionary advantage this way of life has given them. Posing those questions certainly does not diminish my enjoyment of, or wonder at, the natural world. I can readily believe that Richard Dawkins, who has probably pondered those questions more deeply than I have, finds his wonder of nature still further enhanced. Where he and Michael Shermer would part company with Einstein, of course, is in the latter's readiness to use the word "God" in conjunction with this sense of awe and wonder. Perhaps this is partly a consequence of the half century that has elapsed since Einstein's death. "God" has been invoked in that time to justify many things that are, to say the least, morally dubious, including the activities of both sides in the so-called "war on terror". We are sometimes forced to ask if the word "god" can be rescued from the clutches of those who believe in God. Insofar as the criticisms of Dawkins, Shermer and others force believers to confront such questions, they are to be welcomed.

The kind of interaction that Barbour characterized as dialogue is, then, alive and well, as the spate of books that is now appearing on the topic of science and religion testifies. What are we to make of Barbour's fourth mode of interaction, possibly the most controversial of them all, integration? In what sense, if any, can science and religion be integrated? There are many people on both sides who would argue that such a consummation is neither possible nor desirable. Is there a distinction between "integration" and what Stenmark has called "world-view expansionism"?

There are some obviously false ways of attempting "integration". Arguments that science (in this context usually archaeology) has "proved" the Bible, for example, are convincing only to those already converted. Archaeology can certainly help to throw light on the everyday life of biblical times and may even one day, perhaps, help to confirm the historicity of some of the figures, such as King David, who seem in our present state of knowledge to be at least semi-legendary, but it is hardly likely to reveal to us the site of the Garden of Eden, or to unearth Noah's Ark on Mount Ararat.

We also need to resist the temptation into which one twentieth-century Pope fell, to identify modern Big-Bang cosmology with the creation story of the first chapter of *Genesis*. Quite apart from any other considerations, where would that leave biblical exegesis if (as could quite possibly happen) cosmologists move on from today's theory to something that fits the facts even better? Once again Eddington[14], in a slightly different context, made the point concisely: "The religious reader may well be content that I have not offered him a God revealed by the quantum theory, and therefore liable to be swept away in the next scientific revolution." Eddington always strongly opposed that particular kind of integration of science and religion.

I suspect, however, that Eddington would not have objected to the approach taken by the Dalai Lama[15] in his little book *The Universe in a Single Atom*. Rather than trying to show that modern science in some way supports Buddhism, he draws several parallels between scientific ideas and the Buddhist scriptures. His attitude is that both the scientific and the spiritual approaches are needed for a full understanding of our world, and he particularly argues that that is the case if we are to reach a full understanding of the nature of consciousness. We need, he believes, the complementary approaches of the objective studies of neuroscience and the personal studies that can be made by the Buddhist methods of introspection and meditation. He does, however, go on to suggest, particularly in the context of genetic technology, that the whole of society should have a say in what lines of research are pursued. Although his arguments are good ones in the light of the dangers of misuse of certain kinds of knowledge, this is a suggestion that I suspect many scientists will instinctively resist. Those who silenced Galileo, after all, believed that they were acting for the good of society.

Whether or not integration in Barbour's sense is possible, individual scientists will inevitably be influenced by their own religious beliefs (or unbeliefs) in the choices they make among rival theories, or of problems that they wish to study. Stenmark has suggested that this is one way in which individuals will reconcile their own scientific practice and religious beliefs, while Helge Kragh[16] in a recent book, *Matter and Spirit in the Universe: Scientific and Religious Preludes to Modern Cosmology*, has documented how, especially in the field of cosmology, scientists have in fact been guided by their beliefs. Although Kelvin's repeated attacks on the geologists' timescale were based on the scientific arguments that we have already discussed, they were also at least partly motivated by religious beliefs; he felt it important to show that the Earth must have had a beginning. Big-Bang cosmology would have appealed to Kelvin much more than steady-state theories would have done, since the former implies that there was a beginning to the universe, while, as is well-known, one of the motivations for developing the steady-state cosmology was to bring "creation" within the realm of physical science and thus avoid the need to postulate a Creator and a beginning. If, however, a scientist's religious commitment becomes too overt, as many of us feel it has done with those who call themselves "creation scientists", then the results obtained tend to become suspect, and the scientists themselves come to be regarded as what Stenmark called "world-view expansionists". Individual scientists who are, or are perceived to be, committed to proving the action of a Divine Creator, rather than to seeing to what conclusions the facts lead, are likely to find themselves marginalized by the rest of the scientific community and to find that any results they obtain are looked upon with suspicion, or even labelled as "pseudo-science".

One creationist philosopher, Alvin Plantinga[17], wants scientists who are Christians to work within the Christian system of belief which, for him, includes taking account of the possibility of miracles in their scientific work. He calls this kind of science "Augustinian science"; obviously, to speak of "Christian science" in this context would be to invite misunderstanding. Mehdi Golshani[18], an Iranian physicist and philosopher of science, wants Moslem scientists to develop an Islamic science, but there are important differences as well as similarities between what these two men want from their co-religionists. They both argue that it is false to suppose that scientists approach their work without *any* preconceptions, as is sometimes maintained. Many scientists today approach their work with the preconceptions of scientific materialism firmly in their minds. Both Golshani and Plantinga are reacting against this scientific philosophy, which, as I have already pointed out several times, certainly underpins the writings of many of the most widely read popularizers of science today. The two men differ, however, in that Golshani, unlike Plantinga, is quite ready to accept Darwinian evolution and is not concerned to accommodate the miraculous within the purview of science. He wants Moslem scientists to work within the constraints of believing that the universe is a creation, that it has a purpose and moral order, and that there is more to it than we can perceive with our senses. This sounds harmless enough, and is very similar to what I have been urging in this book. Not only scientists who are observant Moslems, but also those who are observant Jews or Christians share those beliefs as, I suspect, do many who are not affiliated with any formal religion. The question is whether the science of such scholars is, can be, or should be, any different from that of their unbelieving colleagues. Golshani, in fact, readily admits that holding these beliefs will make little difference to what most scientists do most of the time. Only those working on what he calls "fundamental theories" (particularly evolution and cosmology), those working on some applications of science, or those engaged in popularizing and interpreting science to the wider public, will find their work directly impacted by such beliefs.

We do have to distinguish between what scientists do in their own research and what they do in their attempts at popularization or, as I would prefer to call it, interpretation. Non-scientists are hungry, not just for titillating accounts of the strange things we encounter in the cosmos, but for serious attempts to relate modern discoveries to our everyday concerns, to what it means to be human, and thus to explain the relevance of those discoveries to our understanding of our own place in that cosmos. Of course, scientists who undertake to offer such explanations and interpretations have to be intellectually honest. If they believe that the scientific-materialist philosophy provides the correct framework for interpretation, they have not only a right, but a duty, to say so; but intellectual honesty also requires that they do not imply that their personal interpretation of the philosophical implications of modern science is the only possible one. Because so many of the most successful of our present-day interpreters of science are scientific materialists, people like Plantinga have come to believe that philosophy to be more characteristic of scientists than it really is. The remedy may be for scientists espousing other points of view to try their hand at the interpretative work of science, rather than to attempt to force their own scientific work into any given set of beliefs, as Plantinga, and to a lesser extent Golshani, would have them do. Religions, after all, are not the only systems that sometimes try to highjack science. Political ideology can do so as well, as is evidenced by the whole sorry story of Lysenkoism in the former Soviet Union, and big business has been known to try to discredit scientific work that adversely affects its immediate interests.

Another form of integration is represented by attempts to find an evolutionary explanation for the origin of religion, such as that made by D.S. Wilson and discussed in Chapter 7. That approach is also an example of what Stenmark calls scientific expansionism, reducing religion to a natural phenomenon. In one sense, these attempts are a mirror image of creationism, which is a form of religious expansionism, although, as I remarked in Chapter 7, they do not necessarily imply that there is no truth content in religions in general. I do not believe, however, that this is what Barbour has in mind when he speaks of integration.

Is there, then, any sense at all in which we can look for integration of science and religion? If there is indeed "one world", as John Polkinghorne[19] had it in the title of one of his books, then science and religion are describing different aspects of it, or, perhaps better, unifying different but overlapping sets of data. The old metaphor of the "two books" is not without merit; there are two very different ways of studying the world we find ourselves in. If the world is one we should, however, be able to obtain a coherent view of it. Perhaps we need to understand both our science and our religion better than we do.

We cannot learn from our science what we should believe in our religion, but we can get some very good pointers to what we should *not* believe. As we saw in Chapter 7, religions were characteristic of the earliest human societies, possibly even of pre-human societies, and thus long antedate the development of modern science; even that of some of the "book religions", to adopt the term Hayden uses for those religions based on scriptures, which appeared before the first attempts of the Greeks and Babylonians (and others) to explore the natural world. It should hardly be a cause for surprise, therefore, that the scriptures of those religions incorporate the commonsense but inadequate (or even incorrect) views that early civilizations entertained. To insist on a literal interpretation of those ideas is not showing respect to the God Whose inerrant word some suppose those scriptures to be, but is, rather, an insult to the God Who gave us the ability to work such things out for ourselves. We can no longer believe that the cosmos revolves around the Earth, still less around the human race. Only by ignoring mountains of evidence (and the evidence of mountains!) can we continue to believe that the Earth is only a few thousand years old and that it, together with the Sun, Moon, stars and planets, was made in six days. Differences of emphasis among evolutionists do not alter the fact that human beings are related to the other animals, and particularly closely related to the giant apes. On the other hand, none of these facts logically rules out the possibility that all this was created for a purpose, or purposes, which is, or are, still obscure to us. It is even possible that the human race has a special role in these purposes, even though it is almost certainly not the central one that we used to like to think it was. Despite clamorous voices raised to the contrary, a rational person can accept all the discoveries of modern science and still believe, but not believe literally, and perhaps not always even traditionally.

For example, the traditional form of the Christian doctrine of original sin, according to which evil and death entered the world when Adam and Eve (whether considered to be literally the first parents of the human race, or simply archetypal figures for the earliest human beings) first disobeyed God can scarcely be reconciled with the suffering inevitably involved in the evolutionary "struggle for existence" over long periods of time before human beings appeared on the Earth – one reason, I suspect, why creationists oppose so strongly both the idea of evolution and the great age assigned to the Earth by modern geology and astronomy. Even before Darwin published *The Origin of Species*, the geology of Hutton and Lyell was sufficient to make the idea of original sin seem much less plausible, and Tennyson was already writing of "nature red in tooth and claw"[20]. Charles Darwin, I believe, was a sensitive and kindly man, and, as we have seen in the last chapter, was well aware of this necessary suffering that came into the world quite independently of any supposed transgression by the earliest members of the human race. His autobiography reveals that his awareness of this was one factor in his slow progression from orthodox belief to agnosticism. Even before that autobiography, he wrote in a letter to the American biologist Asa Gray, specifically of the *ichneumonidae* that lay their eggs in other insects so that the developing larvae can eat their unfortunate hosts from the inside, while the latter are still alive: "I cannot persuade myself that a beneficent & omnipotent God would have designedly created the *Ichneumonidae* with the express intention of their feeding within the living bodies of caterpillars, or that a cat should play with mice."[21] That letter was dated 1860, when, as he also tells us in his autobiography, he still believed in a First Cause – as the rest of the paragraph in the same letter clearly shows. We can, of course, wonder how developed the nervous system of the parasitized insect may be, and to what extent the creature feels the pain of being eaten alive, but Darwin's main point stands: the evolutionary process inevitably causes much suffering. He could see some force in the argument that human suffering helped to improve the moral character of the sufferer, but he could see no point in the suffering imposed on the animal creation. As a result, Darwin first gave up believing in a loving God, and eventually came pretty close to giving up believing in God at all – although it is perhaps not quite certain whether he ended his life as an atheist or an agnostic.

Now, of course, Darwin did not invent the "problem of evil", which has exercised Christian theologians at least since the time of St Augustine and is the central theme of the book *Job* in the Hebrew Bible, but he gave it two new twists: first by drawing attention to non-human suffering and putting it on the same level as human suffering, and second, by making it clear that there was suffering in the world long before human beings came into it and for which, therefore, they could not be held responsible. Carl Sagan made the "cosmic clock" familiar to many people; if the whole history of the universe is represented by a twenty-four hour clock, with the present moment taken as the stroke of midnight, then human civilization has lasted only for the last few seconds and recognizably human ancestors have been around for only a minute or two. That thought expresses graphically what Darwin saw intuitively. Again,

I suspect that it points to one of the underlying motivations of the young-Earth creationists, who do not wish to give up the idea that the human race bears the ultimate responsibility for the presence of evil in the world. It is possible to try to rephrase the doctrine of original sin, as Peacocke[22] does, by pointing to the feeling of being "misfits" in the universe that many feel. We recognize that we are indeed part of the natural world and that our individual deaths are therefore inevitable and yet we feel ourselves to be something more. To people like Bertrand Russell, Steven Weinberg, Richard Dawkins and even Albert Einstein, this feeling is a sign of weakness, a delusion that we must strive to overcome, but it just might be what Wordsworth would have called an "intimation of immortality".

Even if we do so rephrase the doctrine of original sin, however, that hardly makes the human race responsible for the occurrence of evil in the world. A well-known hymn, probably written in Darwin's lifetime although its author died before the publication of *The Origin of Species*, contains the line "every prospect pleases and only man is vile". A thoughtful person can no longer sing that hymn with conviction, for if you look at the prospects closely enough to see the *ichneumonidae* and their victims, the statement is simply untrue. The problem of evil is real enough and Darwin is by no means the only person to have lost his faith over it. In fact, probably the strongest argument that the scientific materialists have is that, if the universe is an entirely naturalistic, perhaps even accidental, purposeless phenomenon, then the strange mixture of good and evil that we encounter within it is exactly what we should expect. If, on the other hand, the universe is the creation of an all-powerful and all-loving God, Who had a definite purpose in creating it, then the existence of any evil at all is a major problem not completely resolved by the traditional argument that to have free will and the risk of suffering is better than to be a happy automaton. The victims of the *ichneumonidae* are not given even that choice.

Natural disasters are not so hard to come to terms with. Their most devastating effects *do* come from our exercise of free will, even though our freedom of will is often exercised in ignorance of the likely consequences. The most devastating effects of these disasters often come from the fact that we choose to build cities on the flood plains of great rivers, in the paths of hurricanes, or on tectonic fault lines which expose the cities to both earthquakes and volcanoes. The death and destruction that occur are the direct result of our own lack of foresight. I myself live near a major fault line and, with my neighbours, talk from time to time of "the big one" that will certainly happen one day. We gamble that it will not happen in our lifetimes. Maybe we will be lucky; if we are not, we should recall some of the last words of Scott of the Antarctic: "We took risks, we knew we took them; things have come out against us, and therefore we have no cause for complaint…"[23] I wrote those words some time before the earthquake that devastated Port-au-Prince and other communities in Haiti. What applies to the comparatively wealthy community in which I live does not necessarily apply to a country in which generations of poverty have resulted in poorly constructed buildings whose collapse increased the death toll. I would not like to be thought to be blaming the people of Haiti for their tragic misfortune, but the whole human race is guilty of the lack of foresight that so terribly increased their troubles.

Physical suffering of the individual that is not the result of his or her choice is, however, once again harder to come to terms with. Of course, we are becoming more and more aware that the lifestyle choices we make can eventually affect our health and happiness, but it remains true that many people have to cope with major illnesses that are no fault of their own. Pain and pleasure, however, are two sides of a coin; the same nervous system that enables us to feel physical pleasure also leaves us vulnerable to physical pain. Consider the difference between being hugged by someone you love and by a bear or, perhaps even worse, a boa constrictor. Presumably – I have not conducted any empirical tests – the initial physical sensations are very similar; the end results, horribly different. Or, again, consider the difference between being tickled for a moment or two in fun and being, quite literally, tickled to death. Pain is a sensory stimulus that either becomes too intense or lasts too long, but which may not be all that different in kind from a stimulus that gives pleasure. We cannot live in this physical world without the risk of pain (and mental suffering largely derives its force from the possibility of physical pain) unless we also give up the possibility of sensual pleasure. The Buddha taught that the root of suffering is in our desires, and one can see what he was getting at; but to learn to live without desiring to be free of pain is not exactly easy. The Christian, in according Divine status to Jesus Christ, who was executed in one of the cruellest ways that human beings have yet devised, is presumably saying that, in some sense, God shares our human suffering, an idea totally opposed to Aristotelian or deist concepts of God and, again, not always easy for those who are suffering to accept, and which also leaves unanswered the question: "If the evil in the world is not ultimately all to be blamed on human disobedience, how did it get there?" Blaming it on "fallen angels" does not help, since this only pushes the problem one stage back.

This problem of the origin of suffering and evil is just one example of the way in which theologians have to take account of the insights of modern science. Both Peacocke, in the book I have just cited, and Polkinghorne[24], in one of his most recent books, *Theology in the Context of Science*, emphasize the need for theologians to include scientific insights in their thinking. A similar conclusion is reached by a somewhat different approach by Michael Heller[25] in his book *The New Physics and a New Theology.* Taking such insights into account is more important, Polkinghorne argues, than being concerned to reconcile the various disputes that have, historically, arisen between scientists and theologians. All three authors lament that not many theologians try to undertake this kind of approach, which may be the only meaningful way in which Barbour's "integration" of science and religion can be realized.

Those theologians who do try to engage seriously with the results of modern science tend to favour one of two answers: process theology and kenotic theology. The first of these, of course, derives from Whitehead's process philosophy and is favoured by Barbour and by those who have been most strongly influenced by him. I confess that I find not only Whitehead, but also most of his would-be interpreters, hard to follow. Barbour[26] summarizes the views of several thinkers of this persuasion arguing that God is the leader of a cosmic community, pre-eminent but not all-powerful. God, in this view, is not "the judge meting out retributive punishment" but "the friend, with us in our suffering and working with us to redeem it". Process theologians, Barbour says, have given up the "traditional expectation of an *absolute victory over evil.*" That would appear to rule out the traditional Christian belief that the decisive battle between good and evil was fought (and won) on the first Good Friday. A struggle is still going on because the process of coming into being is still going on, but God is conceived as always working for good, and able to bring good out of evil. There is clearly some departure from orthodoxy here, but perhaps we should be open to the possibility that our ideas of God should evolve, and some departures from what has hitherto been deemed orthodox are inevitable. A germ of the ideas of process theology can perhaps be discerned in a famous passage in St Paul's *Epistle to the Romans*[27] where he writes (as a modern translation has it): "From the beginning until now, the entire creation, as we know, has been groaning in one great act of giving birth." Or again, the idea of God bringing good out of evil is explicitly stated in the book *Genesis*[28], in the story of Joseph being sold into slavery in Egypt. (If that is seen as the point of the whole story, the question of its historical accuracy pales into insignificance.)

Kenotic theology seems to be preferred on the eastern side of the Atlantic, particularly by those around Polkinghorne and Peacocke. A germ of this kind of theology can also be found in a famous passage from St Paul[29], this time from the *Epistle to the Philippians*, where Christ is spoken of as emptying Himself of His Divine nature, in order to become human. So, in kenotic theology, God has emptied Himself to create a universe according to laws which may, in fact, be statistical at the ultimate level – however convinced Einstein was that God does not play dice. We must suppose (and perhaps this is where faith, or trust, comes in) that all is working to some grand purpose which, when achieved, will transcend all that we experience as evil. On one level, these two theological approaches have much in common, both involving limitations to God's omnipotence as it is commonly understood. In kenotic theology, however, the limitation is self-imposed by God, while in process theology it is imposed upon God – a point that both Polkinghorne and Peacocke stress repeatedly. On the other hand, process theology more obviously takes account of the fact that the universe we live in is, indeed, a process – or so it appears to us. Polkinghorne[30] discusses this point both in his most recent book and in another recent essay, remarking that one possible view (held by Einstein himself) is that the whole of space-time exists eternally. He comes down, however, as I do, on the side of regarding the universe as an unfolding process. One is reminded here of Teilhard de Chardin[31] and his work *The Phenomenon of Man*, which envisaged an evolution of the entire cosmos towards an "omega point". He did his own case a disservice by insisting that his book must be regarded as a scientific treatise, which, of course, it is not. It is, rather, a bold metaphysical and theological speculation, which certainly needs to be revised in the light of scientific knowledge obtained since he wrote it (in the 1930s although it was published posthumously some twenty years later) but which contains insights that, one day, may be more widely recognized.

It is at this point, of course, that atheists and scientific materialists will accuse me of ignoring the evidence. They can certainly point to a large catalogue of evils, both in the past and still going on, to support their case. Many of these evils, too, have been and are being perpetrated in the name of God. Yet the evils of human history cover but a short span of time in one particular location of an immense universe believed to be around 13 billion years old. Recall again the cosmic clock, on the dial of which a whole twenty-four hours represents the entire history of the universe until now, and the human race has been in that universe for less than a minute. Evil for which the human race is directly responsible is a very small part of the whole universe, however malignant it may seem to us. Darwin was nearer to being correct when he saw the evil for which we cannot be held responsible as the chief obstacle to belief in the Christian God. For all we know, however, evil is already decreasing in the cosmos as a whole and St Paul may have been right in suggesting that what we perceive as evil is the birth pangs of creation. Even here on our insignificant Earth, there are signs of progress. However much individuals may fail in their efforts to do right (and I have contributed my share of such failures), moral consciousness *has* emerged, apparently against all the odds. Even within the few millennia of human history we can discern some improvements. Until about two hundred years ago, slavery was accepted as the norm – and a biblically condoned one. We certainly have not eliminated slavery yet, but few people would now defend it. Capital punishment, for murder at least, was also accepted as the norm in most countries until the last few decades; now, many countries have abolished it – the European Union requires abolition as a condition of membership – and even those countries that have retained the death penalty tend to be rather defensive about their practice. On the other side, of course, the twentieth century with its wars and genocides has been one of the bloodiest in history, but that is partly because it was also the century that, so far, has seen the largest number of human beings competing for food and space. A case can be made for a slow and faltering progress towards a greater morality, and, except for the psychopaths among us, we do care about right and wrong and try to do what we believe to be right. If this is happening in human history, which is but a blink of an eye in the cosmic timescale and occupies only a tiny fraction of the universe, what may be happening all over the universe in the billions of years of its history?

In the discussion of the design argument in Chapter 4, we saw that two scientists from quite different disciplines were seriously arguing that life has, in some way, shaped this world, or even the whole universe, to its own needs. We find it counter-intuitive to suppose that something that appeared long after the beginning of the universe could have determined the initial conditions and, indeed, Davies's argument is at present somewhat speculative. But what if life is simply the manifestation, upon this bank and shoal of time, of Arnold's "eternal not ourselves"? Such an idea is quite consistent with Einstein's emphasis on the wonder, or even awe, with which we approach the natural world, and even if Einstein could not take the step that Eddington took of regarding our relationship to the subject of that awe as a personal one, there is nothing to stop *us* from doing so. Those who choose not to take that step should not claim that the facts of science force us to their conclusion.

One of Einstein's objections to the idea of a personal God was the associated idea that God should punish or reward His creatures in an afterlife (in which Einstein also could not believe) that follows the death of the body. Of course, ideas of a Great Assize, or a book in which our every thought and action are recorded, are picture language – imagery again. But if there is a process at work in the universe that includes a tendency towards moral excellence, we can choose, in this life at least, either to work with it or against it. If we choose the latter, we shall be swimming against the tide and encountering difficulties that we may experience as judgement. If we choose to work with that prevailing tendency, we may hope, sooner or later, to be vindicated. I headed the previous chapter with a quotation from Goethe's version of the Faust legend; nearly three centuries before Goethe, Christopher Marlowe[32], in *Doctor Faustus*, treated that legend very differently. Marlowe gave to Mephistopheles a surprisingly modern attitude to Hell. When Faustus asks Mephistopheles why he is not in Hell, where he belongs, the latter replies:

> Why this is hell, nor am I out of it:
> Thinkst thou that I who saw the face of God,
> And tasted the eternal joys of heaven,
> Am not tormented with ten thousand hells
> In being deprived of everlasting bliss!

No-one judged or condemned Mephistopheles who, a few lines earlier, had described himself as one of the

> Unhappy spirits that fell with Lucifer,
> Conspired against our God with Lucifer,
> And are forever damned with Lucifer.

In the ancient story, Lucifer rebelled against God and fell from Heaven, becoming the Devil. Within the story, Lucifer's act was, presumably, a deliberate and free choice, as was also Mephistopheles' decision to join him. Their free choices had inevitable consequences; their torment consisted of being deprived of the bliss they had once known, but that privation was voluntary, not the judgement of a vengeful God. The three Abrahamic faiths all have the concept of judgement, but in Hinduism and Buddhism we find the concept of karma that to some extent takes the place of judgement. To cite the Dalai Lama[33] again, karma implies your actions, and even your thoughts, form your character which, in turn, helps to determine your future thoughts and actions. Few people would dissent from that; indeed, Alison Gopnik says something rather similar in her recent study of childhood development[34]. Karma seems strange to most Westerners only because its influence is seen as reaching from one lifetime into others. We can see the concept of judgement in a similar light: our own free choices do much to determine our fate. We have all done things of which we are ashamed and wish could be undone. One does not have to believe in the literal truth of traditional depictions of Hell to view the prospect of having to give an account of ourselves with some anxiety. If we persist in wrongful activity, it is inevitably destructive of our finer selves, but if we have learned from our mistakes and tried to make amends, we have become that much better and may hope that judgement will be tempered with mercy.

References:

[1] Barbour, I. G., *Religion and Science: Historical and Contemporary Issues* (HarperSanFrancisco, 1997), Chap. 4. (Barbour introduced the terms much earlier, but this is his considered discussion of them.)

[2] Cantor, G. and Kenny, C., *Barbour's Fourfold Way: Problems with his Taxonomy of Science-Religion Relationships*, in *Zygon*, **36**, 765-781, 2001.

3. Stenmark, M., *How to Relate Science and Religion: A Multi-Dimensional Model* (Wm B. Eerdmans Publishing Co., Grand Rapids, Michigan, 2004), pp. 251-253.
4. Gould, S.J., *Rocks of Ages: Science and Religion in the Fullness of Life* (Library of Contemporary Thought: Ballantyne Publishing Group, New York, 1999), p. 5. Pedersen, O., *The Book of Nature* (Vatican Observatory Publications, 1992), (distributed outside Italy by University of Notre Dame Press, Notre Dame, Indiana).
5. Eddington, A.S., *The Nature of the Physical World* (Cambridge University Press, 1928), pp. 350-352, also p. 353.
6. Stanley, M., *Practical Mystic: Religion, Science and A.S. Eddington* (University of Chicago Press, 2007).
7. Brian, D., *Einstein: A Life* (Wiley and Sons, New York, 1996), p. 186 (interview with Einstein).
8. Jammer, M., *Einstein and Religion* (Princeton University Press, 1999), pp. 122-124.
9. Eddington, A.S., *Science and the Unseen World* (George Allen and Unwin Ltd, London, 1929), pp. 49-50.
10. Arnold, M., *Literature and Dogma* (Smith, Elder & Co., London, 1873), Chap. 8, § 1. (Variations on the phrase appear elsewhere in the book).
11. Einstein, A., From *Wie ich in die Welt sehe*, originally published in German. Several slightly different English translations exist. I have used here the version given in Einstein, A., *Ideas and Opinions* (Bonanza Books, Crown Publishers Inc., New York, 1954), p. 11.
12. Dawkins, R., *Unweaving the Rainbow: Science, Delusion and the Appetite for Wonder* (Penguin Books, Harmondsworth, 1998, and Mariner Books, Houghton & Mifflin, New York, 2000), preface.
13. Shermer, M., *Scientific American*, **293**, *Unweaving the Heart*, October, 2005, p. 36.
14. Eddington, A.S., *The Nature of the Physical World* (Cambridge University Press, 1928), p. 353.

15. Tenzin Gyatso (14th Dalai Lama), *The Universe in a Single Atom* (Broadway Books (a division of Random House, New York), 2005), esp. Chaps. 6-9.
16. Kragh, H., *Matter and Spirit in the Universe: Scientific and Religious Preludes to Modern Cosmology* (Imperial College Press, London).
17. Plantinga, A. *Science: Augustinian or Duhemian?* in *Faith and Philosophy*, **13**, 368-394, 1996.
18. Golshani, M., *How to make Sense of Islamic Science?* In *The American Journal of Islamic Studies*, **17**, No. 3, 1-21, 2000.
19. Polkinghorne, J., *One World: The Interaction of Science and Theology* (Princeton University Press, 1986).
20. Tennyson, A., *In Memoriam*, l. 56, 1850.
21. Darwin, C., letter to Asa Gray, May 22nd 1860. Reprinted in *Correspondence of Charles Darwin*, eds. F. Burkhardt *et al.* (Cambridge University Press, 1993), Vol. 8, pp. 293-294.
22. Peacocke, A.R., *Theology for a Scientific Age* (SCM Press, London, 1993), esp. pp. 77, 230, 248-254.
23. Scott, R.F., quoted by Elspeth Huxley in *Scott of the Antarctic* (Weidenfeld and Nicolson, London, 1977), p. 256.
24. Polkinghorne, J., *Theology in the Context of Science* (Yale University Press, New Haven and London, 2009).
25. Heller, M., *The New Physics and a New Theology*, trans. G.V. Coyne (Vatican Observatory and University of Notre Dame Press, Notre Dame, Indiana, 1996).
26. Barbour, I.G., see reference 1, pp. 322-328.
27. *Romans* Chap. 8 vv. 18-25 (the version quoted is that of the *Jerusalem Bible*).
28. *Philippians* Chap. 2 vv. 6-11.
29. *Genesis*, Chap. 50, v. 20.
30. Polkinghorne, J., in *On Space and Time*, S. Majid (ed.) (Cambridge University Press, 2008), pp. 278-283.

31. de Chardin, T., *The Phenomenon of Man*, published in French as *Le Phenomène Humain* (Editions du Seuil, Paris, 1955; English translation, published by William Collins, Sons & Co., London, Harper and Bros., New York, 1959).
32. Marlowe, C., 1604? *The Tragicall History of Dr Faustus*, as given in the *Oxford Dictionary of Quotations*. The speaker is Mephistopheles; ll. 303-5 & 312-6.
33. See reference 15, p. 109.
34. Gopnik, A., *The Philosophical Baby* (Farrar, Strauss and Giroux, New York, 2009), pp. 189-190, 195-196.

EPILOGUE

> We are to admit no more causes of natural things than such as are both true and sufficient to explain their appearances.
>
> Isaac Newton, *Philosphiae Naturalis Principia Mathematica*

In the first chapter of his book *No Free Lunch*, William Dembski[1] makes an impassioned plea for the restoration of design as a possible explanation in science. As he sees it, Francis Bacon and others who are looked on as the founders of modern science turned their backs on design explanations, which have been taboo in science ever since. It is not, of course, quite that simple. Although we rightly look on Bacon and his contemporaries or immediate successors as the founders of what we now call science, they called themselves philosophers – the word *scientist* was not coined until the nineteenth century – and argued only that progress in understanding natural phenomena would best be made by seeking natural causes. Thus Galileo, unlike Aristotle, did not speculate on *why* heavy objects fall to the Earth but concentrated on measuring the rates at which they fell, and so was able to derive the law that the distance travelled is proportional to the square of the time elapsed. In his turn, Newton was able to demonstrate that the inverse-square law of gravitation could explain both that law of the motion of falling bodies and the observed motion of the Moon around the Earth and thus avoided the kind of teleological explanation of the fall of heavy bodies that Aristotle had proposed. That fact, taken with the first of Newton's "Rules of Reasoning in Philosophy", quoted above, certainly looks at first sight as if he were turning his back on any form of design argument; but, as we saw in Chapter 4, this same Newton argued that the solar system could only have been created by the deliberate act of God. As Newton's unpublished writings have become more widely known, it has become increasingly evident that he himself was far from eliminating the purposive intentions of God from his understanding of the world.

Nevertheless, in those areas of study that we now call the natural sciences, the philosophy of reductionism, which many see implied by Newton's rule, has become increasingly popular. The behaviour of living creatures, it is claimed, can be reduced to the workings of molecular biology, which can be reduced to chemistry, which can be reduced to quantum mechanics and the behaviour of fundamental particles. No chemists are quite convinced that everything they study can be predicted from quantum mechanics and, even if that is possible in principle, actually making the predictions eludes us in all but the very simplest cases. It is hard to argue against reductionism as a research programme, because the spectacular success of the natural sciences justifies it as methodology. All scientists are methodological reductionists. My own career in research was spent largely in studying the orbits of stars in binary systems and in trying to learn from them about the sizes, masses and luminosities of stars in general, as well as something of the way in which binary systems evolve. Had I argued that the components of binary systems move and evolve as God intended them to do, however true that might ultimately be, I would have learned very little and not have achieved even the modest degree of recognition from my colleagues that I have received. The success of reductionism as a methodological tool does not necessarily mean, however, that it is the way to explain the meaning of "life, the universe, and everything". Even so, I do not agree with Dembski that we should try to bring design back into science as a possible explanation of phenomena. Rather, we should recognize that the issue of whether or not there is design in the universe is not a matter to be settled by the methods that we have now come to recognize as "scientific" – a conclusion with which Dembski would not agree, since his whole argument is that there are mathematical algorithms that enable us to recognize whether or not design is present.

An alternative view to reductionism is that of *emergence*. Atoms exhibit properties that human beings, at least, could not have predicted from those of the fundamental particles that make them up. Similarly, we cannot predict the properties of molecules from those of their constituent atoms – that is especially true for the large complex molecules we find in living creatures, culminating in the DNA molecule, able to reproduce itself. Still less can we predict the behaviour of living creatures from the properties of their constituent molecules, or the achievements of the human mind from the circuitry of the brain. Thus, it is argued, as matter becomes more organized, potentials are revealed and realized that could never have been predicted from the lower levels of organization. This is a useful corrective to reductionism, but it is still implicitly based on the notion that the fundamental particles of matter are basic and primary; that the universe is a theatre in which life and mind have appeared, whether that was by accident or design. Looking at the universe in that way, we shall probably never reach agreement on that question of accident or design.

For the sake of argument grant for a moment that, despite the spectacular success of methodological reductionism, ontological reductionism is putting the cart before the horse; that life and mind are primary, the manifestations, on this bank and shoal of time, of some greater reality in the realm that I have called the transcendent. It is conceivable that life and mind did not just appear in a pre-existing universe, but that it was they that were pre-existent and shaped the physical universe to their own needs. This is not so very different from what Paul Davies[2] is arguing in *The Goldilocks Enigma*, nor from the belief that God created this enigmatic universe in which we live. If we combine anthropic arguments with the suggestion made by Conway Morris[3] and his associates, that evolution has a goal, namely, the appearance of intelligent beings in this universe, then it is reasonable to infer that the universe was created by a pre-existing Intelligence that, for some reason, willed other intelligences to be manifested in the physical world. As Aquinas might well have added:

"that is what all men call God". Nowadays, many people (men and women) would shy away from using the word "God" even if they accept that argument, which, of course, is no more a "proof" than are Aquinas's own five ways. Both steps in the argument are controversial, but at least that argument demonstrates that modern scientific knowledge does not *necessarily* lead to a materialistic philosophy.

Among Einstein's objections to traditional religions was the notion that any human personality could survive the death of the body. He was, of course, only one of many modern people who find such a prospect incredible. As I remarked in Chapter 8, the traditional Christian imagery in which belief in our survival of death has been clothed no longer seems credible to very many of us. Neither the alleged bliss of Heaven nor the horrors of Hell seem very real to most modern minds. I do not suppose that we can imagine what such a *post-mortem* existence could be really like or even that it is very profitable for us to try to do so but, *if*, as I have suggested in the previous paragraph, mind is in some way prior to the physical universe, the prospect of survival does not seem as absurdly impossible as the scientific materialists try to persuade us that it is. If they are right that death is the annihilation of an individual's personality, that is hardly something to be feared, but each of us has contributed to the sum total of human experience and knowledge and it seems rather wasteful that all that should be snuffed out when the last human being expires. As we have seen, our knowledge of the vastness of the universe has led many to infer that, even if there is a God who created it, such a Deity would have little or no concern for puny creatures like ourselves. Like religious scepticism, that reaction is an old one and can be found, as I have already pointed out, in the eighth Psalm: "When I consider thy heavens, the work of thy fingers, the moon and the stars, which thou hast ordained. What is man that thou art mindful of him? And the son of man that thou visitest him?" Much later, we are assured that even a sparrow does not fall to the ground without God being aware of it, and that we are of more value than many sparrows. If that assurance is true, it is "Good News" indeed.

We live in an enigmatic universe and do not fully understand either the nature of our own minds or the ultimate nature of the matter of which the universe seems to be composed. Indeed, we compound that latter enigma with "dark matter" and "dark energy", which are even more mysterious than their normal counterparts. Such anomalies may be telling us that there are changes in our cosmological theories still to come that will be at least as great as the changes that those theories have undergone already. In the meantime, believers do not have to be literalists, and scientists do not have to be materialists; somewhere we can hope to find a middle ground where we also hope that the truth may lie. Our own generation, the first to have seen pictures obtained both by space probes, including the Hubble telescope, and our large ground-based instruments, has had demonstrated to it that, beyond all doubt, the universe is full of a weird beauty. Traditionally, at least in Western thought, truth, goodness and beauty have always been seen as related. To hope, therefore, that the universe is good as well as beautiful is not "blind faith" based on "no evidence at all"; if we strive to work for the good, at the very least our efforts will not be wasted. In essence, theism is the belief that the power that sustains this enigmatic universe is also the source of life, and that we can have a relationship with that power that is analogous to the most intimate relationships that we enjoy with our fellow human beings. Some people may find that hard to believe, or even unnecessary, but after a lifetime spent in scientific research, I do not know of any result that compels us to abandon the belief.

In the Prologue I argued that each one of us embarked on a quest at the moment of our birth. The quest will certainly continue until we die, and there are perhaps, grounds for hope that it will not end at the time of our death. In any case, others will pursue the quest after us. They will, very probably, build larger telescopes and more powerful particle accelerators, and probe even more deeply than we have done into the molecular basis of life and the processes of evolution. Many aspects of the physical universe that are now obscure to us may well be clarified by such investigations, but, if the past history of science is any guide, the number of present-day problems that will be solved will be less than that of the new problems brought to our attention. I strongly believe that we have a long way to go before we can even hope to find a "theory of everything" and the methods of the natural sciences, valuable though they are, will not answer the questions that are most important to us as human beings. The all-too-obvious faults of institutional religion and the too-literal interpretations that many believers try to force on their scriptures should not blind us to the insights into the human condition that are to be found in the great religions of the world. We do not have to build our philosophy on Bertrand Russell's foundation of "unyielding despair".

References:

[1] Dembski, W., *No Free Lunch* (Rowman and Littlefield, New York, 2002), Chap. 1.

[2] Davies, P.C.W., *The Goldilocks Enigma* (Mariner Books, Houghton Mifflin Co., New York, 2008 (originally published 2006)).

[3] Morris, S. Conway (ed.), *The Deep Structure of Biology: Is Convergence Sufficiently Ubiquitous to Give a Directional Signal?* (Templeton Foundation Press, West Conshohocken, Pennsylvania, 1982).

NAME INDEX

A

Adam 135, 177
Agassiz, L. 157
Alice 91
Alpher, R.A. 3
Anselm, St. 39, 42, 44
Aquinas, St T. 7, 39, 41-44, 46, 50n, 51, 54, 74, 112, 117, 143, 191
Aristotle 5, 7, 43, 54, 112, 191
Arnold, M. 170, 184, 187n
Arrhenius, S. 84, 90n
Augustine, St. 13, 143, 178

B

Barbour, I. G. 165-167, 172-173, 176, 181-182, 186n, 188n
Barnes, E.W. 168.
Barr, J. 39, 50n 118, 129n
Bayes, T. 29-31, 35n, 48, 128, 136
Beethoven, L. van 32
Behe, M. 66-70, 74, 76n
Berkeley, G. 92
Bernstein, L. 32, 36n
Bethe, H. 3
Blakeslee, S. 94, 98, 114n
Bohr, N. 132, 134
Bondi, H. 3
Broad, C.D. 29 35n
Brooks, M. 127, 130n
Browning, E.B. 116, 118

Bruno, G. 6
Buddha, The 27, 28, 110, 181
Burchfield, J.D. 10, 16n
Burnet, T. 151, 163n
Byron, Lord 21

C

Cairns-Smith, A.G. 109, 115n
Cannon, H.G. 154-156, 163n
Carr, B. 60-61, 75n, 133, 148n
Carroll, L. 91
Carter, B. 55, 75n
Chamberlain, T.C. 53
Chapman, A. 151, 163n
Christ (see also Jesus) 80, 88, 119-120, 181-182
Churchland, P. 93-94, 97-98, 113n
Clark, A. 108, 115n
Cleanthes 52
Clerk Maxwell, J. 21
Collins, R. 61, 75n
Conrad, J. 32
Conway Morris, S. 87-88, 90n, 105, 114n, 193, 196n
Copernicus, N. 6,8-9, 16n, 55, 92,150
Copleston, F.C. 41, 50n
Corballis, M.C. 106, 115n
Crick, F.C., vi, xii, xivn, 12, 65, 71, 91, 93-94, 110, 113n, 140, 154, 159

D

Dalai Lama, The 42, 50n, 173, 186, 188n
Dante vi, 5, 6, 15n, 37-39, 46, 145-146, 148n
Darwin, C. v, 7, 10-12, 15n, 18, 32, 55, 64, 65, 68-70, 73, 76n, 83, 92, 102, 123, 130n, 149, 153-154, 159-162, 164n-166, 177-179, 183, 188n
Darwin, E. 153
Darwin, G. 11
David, King 172
Davies, P.C.W. 61-62, 72, 75n, 181, 193, 196n
Dawkins, R. xii, 22, 35n, 44, 47, 71, 76n, 119-120, 124, 126, 129n, 134, 145, 148n, 171-172, 179, 187n
de Chardin, T. 183, 189n
de Morgan, A. 34, 36n
de Vries, H. 12, 16n, 87
Dembski, W. 66-70, 72-73, 76n, 191-192, 196n
Descartes, R. 6, 92, 102
Devil, The (see also Lucifer, Mephistopheles) 145, 149, 186
Diderot, D. 33, 153
Digges, T. 6, 15n
Dodwell, P. 24, 33, 35n, 93, 98, 113n
Dyson, F.W. 56, 75n

E

Eccles, J. 24, 35n, 95, 97, 107-109, 114n, 155-156, 163n
Eckermann, J.P. 149
Eddington, A.S. vi, 1, 14-16n, 33, 36n, 45-50n, 95,109-110, 114n, 122, 161-162, 164n, 167-170, 172-173, 184, 187n
Edwards, P. 111, 115n

Einstein, A. vi, 1, 15n, 32, 40, 66, 142, 162, 165, 168-172, 179, 182-185, 187n, 194
Ellis, G. 60-61, 75n
Elsberry, W.R. 70, 76n
Elsdon-Baker, F. 134, 148n
England, P.C. 11, 16n
Euclid, 40-41
Euler, L. 33-34
Eve, 177

F

Faraday, M. 21-22
Faust/Faustus 149, 185, 189n
Fontenelle, B. le B. 77, 79, 89n
Fossey, D. 102
Frankenstein 21
Freud, S. 55, 123, 130n
Friedmann, A. 3
Frith, C. 95, 114n

G

Galdikas, B. 102
Galilei, G. 5-9, 11, 149-151, 166, 173, 191
Gamow, G. 3, 15n
Geiger, J. 47, 50n, 100, 114n
Gilbert, W. 20
Gingerich, O. 62, 75n
Glashow, S. 133
Gliese, W. 83
God 5-9, 28, 31, 37-39, 42-55, 61-62, 65, 68-70, 73-74, 76n-78, 89n, 99, 114n, 126, 116-121, 123, 129, 131, 135, 137, 140-148n, 150-151, 160-162, 168-172, 177-179, 181-183, 185-186, 192-194
Gödel, K. 43

Goethe, J.W. von 149, 185
Gold, T. 3
Golshani. M. 174-175, 188n
Goodall, J. 102
Gopnik, A. 50n, 186, 189n
Gould, S.J. 87, 90n, 166, 187n
Gray, A. 178, 188n
Gray, P. 32, 36n
Greenfield, S. 92, 103, 113n

H

Haldane, J.B.S. 3, 15n, 17, 22
Hardy, A. 65, 75n, 156, 163n
Harris, S. 38, 44-45, 48-49n
Hawking, S.W. 61
Hayden, B. 125-126, 130n, 176
Heilbron, J.L. 150, 163n
Heisenberg, M. 96, 114n
Heisenberg, W. 109, 137
Heller, M. 181, 188n
Herschel, W. 53, 74n
Hermann, R. 3
Hosea 119, 121, 129n
Hoyle, F. 3, 84, 90n
Hubble, E.P. 1-2, 15n, 34, 55, 195
Hume, D. 51-52, 58, 74n, 92, 109-110, 136, 148n
Hutton, J. 9-10, 64, 152, 157-158, 167, 177
Huxley J 65, 117, 129n, 155
Huxley T 160, 164n

I

Inge, W.R. 77, 88-89n, 168
Isaiah, 119, 121-122, 130n, 132, 134, 148n

J

James, King 9
James, W. 32, 36n, 109-110, 112, 115n
Jeans, J.H. 1, 4, 15n, 53, 167
Jenkins, D. 112
Jesus, (see also Christ) 26, 38, 119-121, 145, 152, 163n, 181
Job 178
John Paul II (Pope) 8
John the Divine, St 121
John, St (Evangelist) 26, 35n
Johnson, G. 87, 90n
Johnson, P. 66
Joshua 8, 10, 151, 167

K

Kant, I. 2, 15n, 22-23, 42, 52-53, 74n
Kauffman, S. 87, 90n
Keats, J. 171
Kekulé, F.A. von S. 121
Kelvin, Lord 10-13, 16n, 158-159, 173
Kenny, A. 41, 42, 50n
Kenny, C. 165, 186n
Kepler, J. 6, 9, 79, 89n, 122, 150
Kierkegaard, S. 119, 129n
King, B. 103, 114n, 116, 126, 129n
Kingsley, C. 160
Kornhuber, H.H. 96, 107
Kragh, H. 7, 15n, 173, 188

L

Lamarck, J-B. 153-157, 163n
Laplace, P-S. Marquis de 29, 52-54, 57-58, 73-74n, 81, 107
Leibniz, G.W. von 44, 109
Lemaître, G. 3
Leslie, J. vi, 57, 60, 75n, 93, 113
Libet, B. 93, 95-98, 101, 107-108, 114n

Lovelock, J. 72, 76n
Lowell, P. 80
Lucifer (see also Devil and Mephistopheles) 185-186
Lyell, C. 9-10, 64, 152, 157-159, 167, 177

M
Marlowe, C. 185, 189n
Mendel, G. 12, 65, 159
Mephistopheles (see also Devil and Lucifer) 185-186, 189n
Michelangelo 32, 135
Miller, K. 69, 70, 74, 76n
Milne, A. A. 47
Milne, E. A. 80, 89n
Mohammed 26, 99, 119
Moloch/Milcom 120
Moore, H. 32
Moro, A-L. 157
Morris, S.C. see Conway Morris, S.
Moses 118-119
Moulton, F.R. 53
Mozart, W.A. 24, 31-34, 36n

N
Newton, I. 6, 9, 32, 37, 41, 52-55, 73-74n, 121-122, 136, 150-151, 171, 191-192
Noah 151, 172
Numbers, R.L. 65, 75n

O
Olbers, H.W.M. 2
Otto, R. 23, 35n
Owen, D. 136, 148n

P
Pascal, B. 1, 8, 63
Paul, St. 39, 99, 113, 119, 141, 148n, 182-183
Peacocke, A.R. 43, 50n, 61, 75n, 112, 115n, 179, 181-183, 188n
Penelhum, T. vi, 143, 148n
Penrose, R. 32, 36n, 93, 108, 113n, 115n
Persinger, M. A. 98-101, 114n
Peters, T. 80, 89n
Philo 52
Pilcher, H. 94, 114n
Pius XII, Pope 118
Plantinga, A. 66, 75n, 174-175, 188n
Plato 24-25, 51, 68, 92, 111
Poincaré, H. 121, 130n
Polanyi, M. 24, 31, 35n, 61, 63, 75n, 122
Polkinghorne, J. 176, 181-183, 188n
Pope, A. 19
Popper, K.R. 23-24, 35n, 95, 106-107, 109-110, 114n, 155, 163n
Pythagoras 111

R
Ramachandran, V.S. 94, 98, 114n
Raven, C.E. 152, 155, 163n, 168
Ray, J. 64, 152, 163n
Rhine, J.B. 29
Rosse, Lord 2
Russell, B. vi, xii-xiv, 32, 36n, 42, 44, 50n, 63-64, 75n, 109-111, 115n, 141, 148n, 179, 196
Rutherford, Lord 11, 33
Ryle, G. 92, 107, 113n

S

Sagan, C. xiii, 178
Sayers, D.L. vi, 145-146, 148n
Scopes, J.T. 12
Scott, E 66, 76n
Scott, R F. 180, 188n
Shakespeare, W. 14, 22, 35n, 91
Shelley, M.W. 21-22, 35n
Shermer, M. 31, 35n, 171-172, 187n
Shiva 89
Simpson, G.G. 87, 90n
Skinner, B.F. 102
Soal, S.G. 29
Socrates 26, 113
Spinoza, B. 109, 168
Stenmark, M. 125, 130n, 165-167, 172-174, 176, 187n
Stevenson, I. 111-112, 115n
Stubenberg, L. 109-110, 115n
Stumpf, E. 143, 148n
Swinburne, R. 93, 97, 113, 136, 143, 148n

T

Tarter, J. 20, 35n
Tempier, Bishop 7
Tennyson, Lord 131, 177, 188n
Thomson, W. see Kelvin, Lord
Tryon, E.P. 43, 50n
Turner, J. S. 71-72, 76n, 156, 163n
Tyndall, J. 142-143, 148n

U

Ussher, J. 9

V

Virgin Mary, 118

W

Waltham, D. 83, 89n
Watson, J. N. 12, 65, 159
Wells, H.G. 25-26, 35n, 80, 89n
Werner, A.G. 157
West, D.J. 138-139, 141, 148n
Wheeler, J.A. 61
Whitehead, A.N. 167, 181
Wickramasinghe, N.C. 84, 90n
Wilson, D.S. 124-126, 130n, 176
Wilson, E.O. 123-124, 130n, 133, 148n
Wordsworth, W. ix, x, xii, 26, 179

X

Xenophanes 127

Y

Young ,T. 62

"n" signifies that the entry is a bibliographic note. Names that appear *only* in the Preface or in bibliographic notes have not been indexed.

SUBJECT INDEX

A
Angels 78-79, 132, 134, 181
Anthropic principle 56, 59-60, 63, 66
Atheism 53, 127, 153, 155, 160

B
Bayes Theorem 29-31, 48 128, 136
Bayesian statistics 58, 63, 87, 128
Bible 8-9, 117, 119-120, 136, 151-153, 158, 166, 172, 178
Biblical criticism 152, 161
Biblical literalism/literalists 10, 69, 117, 152, 167
Brain ix, xii-xiii, 21, 26, 38, 47, 69-71, 91-102, 104-105, 107-108. 110, 112-113, 127-128, 140, 155, 193
Buddhism/Buddhist 13, 38, 42, 48, 61, 88, 111, 147, 173, 186

C
Christianity/Christians x, 7, 13, 28, 31, 38, 47-48, 54-55, 64, 67-68, 71, 78, 88, 89, 111-113, , 117, 119-121, 129, 135, 155, 167, 174-175
Consciousness 17, 38, 56, 72, 77, 93, 98, 110-111, 116, 169-170, 173, 184
Creation, artistic 24, 31-33, 122-123, 149
Creation, The 6, 7, 9, 53, 66, 78, 118, 127, 135, 152-153, 172, 174, 179, 182, 184
Creationism/creationist 9, 65-66, 87, 144, 157, 161, 174, 176-177, 179
Creeds 48, 160

D
Darwinism 11-12, 65-66, 70, 154-156, 159-160,
Death 38, 98, 101, 107, 110-111, 168, 177, 179, 185, 194-195
Deism/Deist 54, 61, 141-142, 151, 153, 158, 160, 181
Design, Argument from 43, 51-74, 161, 184, 191-193
Design, Intelligent 52, 66-68, 70-71, 74, 155, 159, 161
Discovery, scientific xiv, 24, 31-33, 121-123
DNA 12, 65, 72, 84-85, 154-155, 159, 161, 193
Dualism 92, 98, 107-108, 110, 112-113

E
Evolution (biological) 10-13, 18-19, 25, 52, 64-70, 72, 83, 87-88, 105-106, 126, 144, 153-159, 161-162, 167, 174-175, 177, 183, 193, 195

F
Faith 44-49, 118, 128, 141, 149, 166, 182, 195
Fine-Tuning 61-62, 133

G
Genes x, 124, 134, 155, 157
Genetics 12, 65-66, 92, 155, 159
Geology v, 10, 151, 157, 159, 167, 177

H
Heaven 22, 32, 78, 116, 118, 127, 150, 185-186, 194
Hell 145-146, 150, 185-186, 194
Hindu/Hinduism 13, 61, 88-89, 111, 129, 186

I
Imagery, in religion 6, 28, 78, 113, 117, 131-135, 147, 170, 185, 194
in science 78, 132-134, 142
Intelligence 63, 91, 105, 193
Islam/Moslem xi, 28, 38, 47, 55, 99, 129, 135, 174-175,

J
Jews/Judaism 38, 47, 55, 135, 168, 175

L
Lamarckism 155
Life, beyond Earth 20, 78-85, 88-89, 91
origin of 13, 56-63, 72, 83-85, 88, 133, 192-193, 195
Lourdes 31, 137-141, 146

M
Materialism xii, xiv, 22, 25, 29, 37, 92, 95, 100, 109-111, 155, 163, 174
Mind 24, 70, 72, 92-97, 102-103, 107-112, 137, 140-141, 193-194
Miracles 31, 131, 135-141, 174
Multiverse 58, 60-61, 133
Music 31-33, 107

N
Natural selection 10, 12, 18-19, 64-71, 74, 87, 104, 124, 155-159, 161
Neuroscience v, 96, 99, 173
Near-death experience 100-101
Neutral Monism 109-111

O
Original sin 177, 179

P
Pamspermia 84-85
Pantheism 168
Planetary systems vi, 6, 81
Prayer 131, 141-147, 150
Proofs (of God's existence) 39-43, 46, 51, 73-74, 194
Purpose (in the universe) 8, 14, 55, 64, 77, 166, 175, 177, 179, 182

Q
Quantum Theory 14, 58, 60, 109, 161-162, 172
Qur'an 88-89, 99

R

Reductionism 94, 96, 192-193
Reincarnation 111-112
Relativity 1, 66, 109-110, 161-163
Revelation 28, 33, 43, 51, 117-123, 128
Rig Veda 127

T

Theology, Kenotic 181-183
Theology, Process 181-183
Third-man factor 47, 100, 147
Transcendent 23-29, 31, 34, 37-38, 48-49, 78, 109-110, 117, 119, 123, 128, 193

MELROSE BOOKS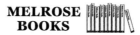

If you enjoyed this book you may also like:

Essays of the Sky
Cosmo Laurence

The term Space Age is defined very concisely in dictionaries but its full meaning takes in great territory. *Essays of the Sky*, cover-to-cover, examines that territory.

It goes where mainstream literature seldom ventures: into such space-relevant topics as the limits of science, the conundrums of religion, the flights of philosophy, the walls of bureaucracy, and the refusal of a rumbling world to rationally face a teeming vastness.

Ruthless rigor, biting satire, and instructive drama combine to penetrate thick taboos, rendering the Paranormal plausible, the reach of God imaginable, the Extraterrestrial respectable, a chorus of knee-jerk denials moot, and the Bible space-age modern.

Prepare, then, for an expansive journey beyond the warm-and-fuzzy orthodox: for a free flow of the universal in an age that by reputation has discovered a Universe. Read of a cosmic model superior to either a Big Bang or a Steady State, of why "parallel universes" amount to nonsense, and of how cosmic structure could ensure an eternity of thriving planets, suns, moons, and galaxies. Process bold, new definitions of common words: time, space, information, superstition, randomness, intelligence, and miracle. Take a clean break from prevailing dogma for a fresh understanding of the greater environment.

Size: 234mm x 156mm Pages: 333
Binding: Paperback ISBN: 978-1-907040-53-5 £13.99

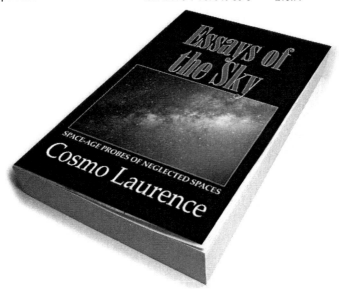

St Thomas' Place, Ely, Cambridgeshire CB7 4GG, UK
www.melrosebooks.com sales@melrosebooks.com